Machine Learning in Translation Corpora Processing

Krzysztof Wołk

Department of Multimedia
Polish-Japanese Academy of Information Technology
Warsaw, Poland

CRC Press
Taylor & Francis Group
Boca Raton London New York

CRC Press is an imprint of the
Taylor & Francis Group, an **informa** business

A SCIENCE PUBLISHERS BOOK

CRC Press
Taylor & Francis Group
6000 Broken Sound Parkway NW, Suite 300
Boca Raton, FL 33487-2742

First issued in paperback 2020

ISBN-13: 978-0-367-18673-9 (hbk)
ISBN-13: 978-0-367-78020-3 (pbk)

Visit the Taylor & Francis Web site at
http://www.taylorandfrancis.com

and the CRC Press Web site at
http://www.crcpress.com

Acknowledgements

First of all, I want to thank everyone, without whom this work would not be accomplished. Especially, much gratitude goes to my family—mostly to my wife Agnieszka, for her help, patience and support. Additionally, I dedicate this diploma to my children, Kamil, Jagoda and Zuzanna, because they provided me with motivation and time for this work.

Special thanks go to my parents, for their effort in raising me and having faith in me and in my actions. Without them, I would not be in the place where I am now.

Lastly, I thank my supervisor and mentor, Professor Krzysztof Marasek, for all the help and hard work he gave to support me—especially for his valuable comments and support in my moments of doubt.

Preface

The main problem investigated here was how to improve statistical machine speech translation between Polish and English. While excellent statistical translation systems exist for many popular languages, it is fair to say that the development of such systems for Polish and English has been limited. Research has been conducted mostly on dictionary-based [145], rule-based [146], and syntax-based [147] machine translation techniques. The most popular methodologies and tools are not well-suited for the Polish language and therefore require adaptation. Polish language resources are lacking in parallel and monolingual data. Therefore, the main objective of the present study was to develop an automatic and robust Polish-to-English translation system in order to meet specific translation requirements and to develop bilingual textual resources by mining comparable corpora.

Experiments were conducted mostly on casual human speech, consisting of lectures [15], movie subtitles [14], European Parliament proceedings [36] and European Medicines Agency texts [35]. The aims were to rigorously analyze the problems and to improve the quality of baseline systems, i.e., adaptation of techniques and training parameters, in order to increase the Bilingual Evaluation Understudy (BLEU) [27] score for maximum performance. A further aim was to create additional bilingual and monolingual data resources by using available online data and by obtaining and mining comparable corpora for parallel sentence pairs. For this task, a methodology employing a Support Vector Machine and the Needleman-Wunsch algorithm was used [19], along with a chain of specialized tools. The resulting data was used to enrich translation systems information. It was adapted to specific text domains by linear interpolation [82] and Modified Moore-Lewis filtering [60]. Another indirect goal was to analyze available data and improve its quality, thus enhancing the quality of machine language system output. A specialized tool for aligning and filtering parallel corpora was developed.

This work was innovative for the Polish language, given the paucity of good quality data and machine translation systems available, and meets pressing translation requirements. While great progress has been made, it is fair to say that further research needs to be conducted in order to refine the system. However, the present study shows great potential in helping people accurately translate a difficult language with a minimum of errors. The improvements in translation were evaluated with success during IWSLT[1] evaluation campaigns. The Wikipedia and Euronews portals were successfully mined for bi-sentences. Improvements in machine translation systems through the use of comparable corpora were demonstrated. Additional improvements to the BLEU metric were also made in order to adapt it to the needs of the Polish language.

In addition, methodologies to facilitate automatic evaluation metrics in other fields of research, methods for generating virtual parallel corpora and parallel data preprocessing are shown as well. A variation of BLEU metric and implementation of HMEANT metric for Polish language are introduced.

---

[1] www.iwslt.org

Contents

Abbreviations and Definitions

Bilingual Evaluation Understudy (BLEU): An algorithm for evaluating the quality of text that has been machine-translated from one natural language to another. Scores lower than 15 mean that the machine translation engine was unable to provide satisfactory quality, as reported by Lavie [19] and a commercial software manufacturer [18]. A high level of post-editing will be required in order to finalize output translations and attain publishable quality. A system score greater than 30% means that translations should be understandable without problems. Scores over 50% reflect good, fluent translations.

Compound splitting: When dealing with morphologically-rich languages, a splitting step is necessary to produce correct translation of compound terms into other languages.

Corpus: A corpus is a collection of text stored on a computer. There are two types:

- **Comparable corpus:** This consists of texts in two languages that are similar in content but are not translations of each other.
- **Parallel corpus:** A parallel corpus contains texts in two languages and is typically used to refer to texts that are translations of each other, aligned at least at sentence level.

EuroParl: European Parliament proceedings.

HAMT: Human-Aided Machine Translation, where the output is post-edited by a human in order to upgrade the quality of the computer output.

IWSLT: International Workshop on Spoken Language Translation.

Moses: A statistical machine translation system that allows automatic training of translation models for any language pair in a parallel corpus.

MT: Machine Translation.

METEOR: Metric for Evaluation of Translation with Explicit Ordering, a method that takes into account several factors that are not considered in BLEU.

SMT: Statistical Machine Translation.

SLT: Spoken Language Translation.

TED: Technology, Entertainment and Design is a forum in which invited lecturers discuss important topics in science, business and global issues that are translated into more than 100 languages. Those corpora are a great example of spoken language.

SVO: An abbreviation used to represent texts with words reordered to match Subject-Verb-Object order.

GIZA: A tool used for alignment of bilingual texts.

MERT: A method used in Moses for translation parameters tuning.

SRILM: A tool used for language model estimation.

MML: An abbreviation used to represent texts that were filtered using Modified Moore-Lewis filtering.

NW: An abbreviation for the Needleman-Wunsch algorithm.

A*: An abbreviation for the A* Search algorithm.

EUP: European Parliament Proceedings.

OPEN: An abbreviation used for movie subtitles.

EMEA: European Medicines Agency.

ASR: Automatic Speech Recognition.

TTS: Text to Speech.

S2S: Speech to Speech translation.

AR: Augmented Reality.

OCR: Optical Character Recognition.

POS: Part-of-Speech tagged texts.

LM: Language Model.

SVM: Support Vector Machines.

OOV: Out-of-Vocabulary word.

PL: An abbreviation used for the Polish language.

EN: An abbreviation used for the English language.

Overview

This monograph is structured as follows: Chapter 1 is an introduction that provides information about aim, scope and motivation for this monograph. It also gives some background information about the speech translation problem for the Polish language and introduces the issue with some basic terms. Chapter 2 provides technical background on the problems and techniques used in statistical machine translation and in comparable corpora exploration. The tools used in this research, as well as baseline configuration settings of the translation systems, are described in this chapter. In addition, this chapter mentions methods for machine translation evaluation. Chapter 3 explains the current state of the art in Polish-English translation and in comparable corpora mining. It also discusses the Polish (PL) language in comparison to other world languages and progress made (based on the IWSLT Evaluation Campaigns in 2013 and 2014). In Chapter 4, the author's solutions to problems outlined in previous chapters and improvements to current methods are described. Implementation of the author's solutions to problems encountered during this research is also explained. A parallel data mining method for comparable corpora is enhanced, a parallel data filtering and aligning tool is introduced and the BLEU [27] metric is adapted to the specific needs of PL-EN speech translation. In addition, experiments designed by the author to evaluate the solutions are described. Chapter 5 presents and discusses evaluation results for the SMT, comparable corpora mining, filtering and aligning tool and enhanced BLEU metric [27]. Conclusions are drawn, based on the results, and briefly summarized in Chapter 6.

1

Introduction

The aim of this monograph is to develop, implement, and adapt methods of statistical machine translation (SMT) [85] to Polish-English speech translation requirements. During a conversation, real-time speech translation would allow the utterances to be immediately translated and read aloud in another language, provided that the system is connected with text-to-speech (TTS) [151] and automatic speech recognition (ASR) [151] systems. Speech translation enables speakers of different languages to communicate freely. It also has great importance in the field of science, as well as in intercultural and global data exchange. From a business point of view, such speech translation systems could be applied as an aid to simultaneous interpreting, as well as in the field of respeaking.

Another aspect of the study is the preparation of parallel and comparable corpora and language models enhanced by pre- and post-processing of the data using morphosyntactic analysis. Such analysis enables the changing of words into their basic forms (to reduce the vocabulary) and the standardizing of the natural order of a sentence (especially to Subject Verb Object [SVO] word order [150], in which the subject is placed before the predicate and the object is located at the end). Training a factored translation model enriched with part of speech (POS) tags [85] was also addressed.

This study improves SMT quality through the processing and filtering of parallel corpora and through the extraction of additional data from the resulting comparable corpora. In order to enrich the language resources of SMT systems, various adaptation and interpolation techniques were applied to the prepared data. Experiments were conducted using spoken data from a specific domain (European Parliament proceedings and written medical texts [35, 36]), from a wide domain (TED lectures on various topics [15]) and from human speech (on the basis of movie and TV series dialogs [14]).

Evaluation of SMT systems was performed on random samples of parallel data using automated algorithms to evaluate the quality [149] and potential usability of the SMT systems' output.

As far as experiments are concerned, the Moses Statistical Machine Translation Toolkit software [8], as well as related tools and unique implementations of processing scripts for the Polish language, were used.

Moreover, the multi-threaded implementation of the GIZA++ [148] tool was employed in order to train models on parallel data and to perform symmetrization at the phrase level. The SMT system was tuned using the Minimum Error Rate Training (MERT) tool [45], which, through parallel data, specifies the optimum weights for the trained models, improving the resulting translations. The statistical language models from single-language data were trained and smoothed using the SRI Language Modeling (SRILM) toolkit [9]. In addition, data from outside the thematic domain was adapted. In the case of parallel models, Modified Moore-Lewis filtering (MML) [60] was used, while single-language models were linearly interpolated. In order to enrich the training data, morphosyntactic processing tools were employed. (The morphosyntactic toolchain created by The WrocUT Language Technology Group and the PSI-Toolkit created by Adam Mickiewicz University [59] were used.)

Lastly, the author developed and implemented a method inspired by the Yalign [19] (parallel data-mining tool). Its speed was increased by reimplementing Yalign's algorithms in a multi-threaded manner and by employing graphics processing unit (GPU) computing power for the calculations. The mining quality was improved by using the Needleman-Wunsch [20] algorithm for sequence comparison and by developing a tuning script that adjusts mining parameters to specific domain requirements.

1.1 Background and context

Polish is one of the more complex West-Slavic languages, in part because its grammar has complicated rules and elements, but also because of its large vocabulary, which is much larger than that of English. This complexity greatly affects the data and data structures required for statistical models of translation. The lack of available and appropriate resources required for data input to SMT systems presents another problem. SMT systems work best in specified text domains (not too wide) and do not perform well in general purpose use. High-quality parallel data, especially in a required domain, has low availability. All those differences and the fact that Polish and West-Slavic group has been, to some extend, neglected in this field of research, makes PL-EN translation an interesting topic of research as far as translation and additional resourcing is concerned. PL-EN results should be repeatable also for other languages in the West-Slavic group.

In general, Polish and English also differ in syntax and grammar. English is a positional language, which means that the syntactic order (the order of words in a sentence) plays a very important role, particularly due to the limited inflection of words (e.g., lack of declension endings). Sometimes, the position of a word in a sentence is the only indicator of the sentence's meaning. In a Polish sentence, a thought can be expressed using several different word orderings, which is not possible in English. For example, the sentence "I bought myself a new car" can be written in Polish as "Kupiłem sobie nowy samochód," or "Nowy samochód sobie kupiłem," or "Sobie kupiłem nowy samochód," or "Samochód nowy sobie kupiłem." The only exception is when the subject and the object are in the same clause and the context is the only indication of which is the object and which is subject. For example, "Pies liże kość (A dog is licking a bone)" and "Kość liże pies (A bone is licking a dog)."

Differences in potential sentence word order make the translation process more complex, especially when using a phrase-model with no additional lexical information [18]. Furthermore, in Polish it is not necessary to use the operator, because the Polish form of a verb always contains information about the subject of a sentence. For example, the sentence "On jutro jedzie na wakacje" is equivalent to the Polish "Jutro jedzie na wakacje" and would be translated as "He is going on vacation tomorrow" [160].

In the Polish language, the plural formation is not made by adding the letter "s" as a suffix to a word, but rather each word has its own plural variant (e.g., "pies-psy", "artysta-artyści", etc.). Additionally, prefixes before nouns like "a", "an", "the", do not exist in Polish (e.g., "a cat-kot", "an apple-jabłko", etc.) [160].

The Polish language has only three tenses (present, past, and future). However, it must be noted that the only indication whether an action has ended is an aspect. For example, "Robiłem pranie" Would be translated as "I have been doing laundry", but "Zrobiłem pranie" as "I have done laundry", or "płakać-wypłakać" as "cry-cry out" [160].

The gender of a noun in English does not have any effect on the form of a verb, but it does in Polish. For example, "Zrobił to. – He has done it", "Zrobiła to. – She has done it", "lekarz/lekarka - doctor", "uczeń/uczennica = student", etc. [160].

As a result of this complexity, progress in the development of SMT systems for Polish has been substantially slower than for other languages. On the other hand, excellent translation systems have been developed for many popular languages. Because of the similar language structure, the Czech language is, to some extent, comparable to Polish. In [161], the authors present a phrase-based machine translation system that outperformed any previous systems for Wall Street Journal text translation. The authors were able to achieve BLEU scores as high as 41 by applying various techniques for improving machine translation quality. The authors adapted an alignment symmetrization method to the needs of the Czech language and used the lemmatized forms of words. They proposed a post-processing step for correct translation of numbers and enhanced parallel corpora quality and quantity.

In [162], the author prepared a factored, phrase-based translation system from Czech to English and an experimental system using tree-based transfer at a deep syntactic layer. In [163], the authors presented another factored transition model that uses multiple factors taken from annotation of input and output tokens. Experiments were conducted on news texts that were previously lemmatized and tagged. A similar system, when it comes to text domain and the main ideas, are also presented for Russian [164].

Some interesting translation systems to and from Czech to languages other than English (German, French, Spanish) were recently presented in [165]. The specific languages properties were described and their properties were exploited in order to avoid language-specific errors. In addition, an experiment was conducted on translation using English as a pivot language, because of the great difference in parallel data.

A comparison of how the relatedness of two languages influences the performance of statistical machine translation is described in [166]. The comparison was made between the Czech, English, and Russian languages. The authors proved that translation between related languages provides better translations. They also concluded that, when dealing with closely-related languages, machine translation is improved if systems are enriched with morphological tags, especially for morphologically-rich languages [166].

Such systems are urgently required for many purposes, including web, medical text and international translation services, for example, for the error-free, real-time translation of European Parliament Proceedings (EUP) [36].

1.1.1 The concept of cohesion

Cohesion refers to non-structural resources, such as grammatical and lexical relationships in discourse. Cohesion is increased by the ties that link a text together and make it meaningful, for which the purpose and quality characteristics are not usually known or seen as relevant.

It is difficult to avoid using human subjects in evaluations of assimilative translation [95], in which the goal is to provide a translation that is good enough to enable a user with little knowledge of the source language to gain a correct understanding of the contents of the source text. In [95], Weiss and Ahrenberg present two methods to deal with aspect assignment in a prototype of a knowledge-based English-Polish machine translation (MT) system. Although there is no agreement among linguists as to its precise definition (e.g., Dowty [96]), aspect is a result of the complex interplay of semantics, tense, mood and pragmatics. It strongly affects the overall understanding of the text. In English, aspect is usually not explicitly indicated on a verb. On the other hand, in Polish it is overtly manifested and incorporated into verb morphology. This difference between the two languages makes English-to-Polish translation particularly difficult, as it requires contextual and semantic analysis of the English input to derive an aspect value for the Polish output [97].

1.2 Machine translation (MT)

Next, the history, approach, and applications and research trends of MT systems will be discussed.

1.2.1 History of statistical machine translation (SMT)

In recent years, SMT has been adopted as the main method employed for machine translation between languages. The main reasons for the use of SMT are its accuracy and the fact that it does not require manually-constructed rules [152]. A number of translation engines have been developed for general translation purposes, with perhaps the best known being Google Translate, which is available on the Internet for non-commercial purposes [133].

The enormous demand for translation services for different languages in science and technology has nearly always exceeded the capacity of translation professionals, i.e., actual humans. This work was driven, in part, by the universal availability of the Internet, where a user can access content in virtually any language. It was impossible for the huge demand for instant translation to be met by human translators on such a massive scale. Thus, much effort was expended in order to develop MT software products specifically for translating web pages, textbooks, European Parliament and something as mundane as—but still very useful—email messages.

During the 1980s, the forerunner of immediate online translation services was the rule-based Systran system, which originated in France and was limited to the Minitel network [134]. By the mid-1990s, many MT providers were offering Internet-based, on-the-spot translation services. In subsequent years, machine translation technology has advanced immeasurably, with Google translate supporting fifty-seven languages [135]. However, the overall translation quality of online MT services is frequently poor when it comes to languages other than English, French, German and Spanish. On the other hand, improvements are continuously being made. Google and other developers offer an automatic translation browsing tool, which can translate a website into a language of your choice [134]. While these services provide acceptable, immediate, "rough" translations of content into the user's own language, a particular challenge for MT is the online translation of impure language that is colloquial, incoherent, not grammatically correct, full of acronyms and abbreviations, etc. For obvious reasons, when scientific, medical and technical texts are being translated, the accuracy of the MT is of paramount importance. In general, the use of systems with satisfying quality for a narrow domain (e.g., medical texts) is quite possible.

The rapid translation of huge amounts of text is a very repetitive and tedious task that is ideally suited to computers. Advantage can be taken of the fact that much of the translated documentation does not vary greatly from one version to another. This led to the creation of translation memories (for example, Directorate-General for Translation [DGT] translation memory), which became a core technology in MT; they enable previous translations to be easily retrieved and automatically inserted into new translation projects [167].

According to Jean Senellart [92], machine translation is following technological trends in the speech area at a slower rate than speech recognition is. In addition, some ideas from the speech recognition research field were adopted for statistical machine translation. This interdisciplinary collaboration between researchers led to the establishment of many projects, like the Global Autonomous Language Exploitation (GALE) program, funded by the U.S. Defense Advanced Research Projects Agency (DARPA), and the Technology and Corpora for Speech to Speech Translation (TC-STAR) [199, 200] program, funded by the European Commission (2004–2007) and the Asia Speech Translation Consortium (A-STAR) [204]. Through such programs, knowledge is transferred between the two disciplines.

1.2.2 *Statistical machine translation approach*

In most cases, there are two steps required for SMT: An initial investment that significantly increases the quality at a limited cost, and an ongoing investment to incrementally increase quality. While rule-based MT brings companies to the quality threshold and beyond, the quality improvement process may be long and expensive [137]. Rule-based machine translation relies on numerous built-in linguistic rules for each language pair. Software parses text and creates a transitional representation from which the text in the target language is generated. This process requires extensive lexicons with morphological, syntactic and semantic information and large sets of rules. The software uses these complex rule sets and then transfers the grammatical structure of the source language into the target language. Translations are built on gigantic dictionaries and sophisticated linguistic rules. Users can improve out-of-the-box translation quality by adding their own terminology into the translation process. They create user-defined dictionaries, which override the system's default settings [137]. The rule-based translation (RBMT) process is more complex than just substitution of one word for another. For such systems, it is necessary to prepare linguistic rules that would allow such words to be put in different places and to change their meaning depending on context, etc. The RBMT methodology must apply a set of linguistic rules in three phases: Syntax and semantic analysis, transfer, syntax and semantic generation [168].

An MT system, when given a source text, first tries to segment it, for example, by expanding elisions or marking set phrases. The system then searches for such segments in a dictionary. If a segment is found, the search will return the base form and accompanying tags for all matches (using morphological analysis). Next, the system tries to resolve ambiguous segments (e.g., terms that have more than one match) by choosing only one. RBMT systems sometimes add an extra lexical selection that would allow choices between alternative meanings.

One of the main problems with the RBMT approach to translation is making a choice between correct meanings. This involves a disambiguation, or classification, problem. For example, to improve accuracy, it is possible to disambiguate the meanings of single words. The biggest problem with such systems is the fact that the construction of rules requires a great effort. The same goes for any sort of modifications to the rules that do not necessarily improve overall system accuracy.

At the most basic level, a MT system simply substitutes words in one language for words in another. However, this process cannot produce a good translation of a text, because recognition of whole phrases and their closest counterparts in the target language is needed. Solving this problem with corpus and statistical techniques is a rapidly growing field that is leading to better translations [85].

SMT approaches the translation of natural language as a machine-learning problem and is characterized by the use of machine learning methods. By investigating and examining many examples of human-produced translation, SMT algorithms automatically learn how to translate. This means that a learning algorithm is applied to a large body of previously-translated text, known as a parallel corpus. The "learned" system is then able to translate previously-unseen sentences. With an SMT toolkit and enough parallel text, a MT system for a new language pair can be developed in a short period of time. The correctness of these systems depends significantly on the quantity, quality and domain of the available data.

1.2.3 SMT applications and research trends

Currently, there are many potential applications for statistical machine translation. First, SMT can be integrated with speech technologies. Such a combination with speech would open up a broad range of possible applications, such as the translation of phone conversations or any other broadcasts in real-time. Recently, Skype voice communicator started tests of such translations.[2] Because methods used in modern speech recognition and statistical machine translation are quite similar or, in some cases, even identical, the integration of the two is straightforward. In the simplest possible application, the output of a speech recognition system can be passed directly to an SMT system. In more complex scenarios, since both technologies operate on statistical models, word lattices with confidence scores can also be exchanged [169].

In addition, a rapid increase in the computing power of tablets and smartphones, together with wide availability of high-speed mobile Internet access, makes it possible to run machine translation systems on them (of course, only if such systems are scaled down to match the device memory limitations). Such experimental systems have been developed in order to assist foreign health workers in developing countries. Similar systems are already available on the commercial market. For example, Apple's iOS 8 and above allows users to dictate text messages. Initially, a built-in ASR system recognizes the speech. In the second step (if Internet access is available), the recognition results are altered by an online system. Supplying it with SMT for text and voice message translation seems to be a natural next step. Another potential commercial use that comes to mind is tourism and health care services for foreigners.

Mobile text translators are already widely available and commercially successful. The translation of full sentences recognized from spoken language is on the horizon, because it is already technically possible. Projects such as Universal Speech Translation Advanced Research (U-STAR,[3] a continuation of the A-STAR project) and EU-BRIDGE[4] are currently conducting research in this area. Increasing interest in combining speech recognition, machine translation and speech synmonograph has been seen in recent years [93]. To achieve speech-to-speech translation, often n-best lists are passed from the ASR to the SMT system. Unfortunately, combining those systems raises problems—such as sentence segmentation, de-normalization and punctuation prediction—that must be solved in order to achieve good quality translations [94].

---

[2] http://www.skype.com/en/translator-preview/
[3] http://www.ustar-consortium.com/
[4] http://www.eu-bridge.eu/

From an economic perspective, MT is a timesaving solution, and the lack of accuracy can be countered by human-aided machine translation (HAMT) [136], where the computer output is post-edited by a human who upgrades the output to an acceptable publishable quality, i.e., "MT for dissemination."

Another possibility for the development of mobile speech technologies is an extension for translators that would supply them with data obtained from an optical character recognition (OCR) system [153]. The images taken from the camera of a mobile device could be translated on a mobile device or even replaced using augmented reality (AR) [153] capabilities in order to aid tourists and other users. Such a solution would allow foreigners to translate road signs, directions, instructions, warnings in public areas, restaurant menus, documents, etc. Adding speech synmonograph to this solution would allow deaf people to "read" important information in their language using hand-held devices. In an application like this, the text domain is very limited, so only a small, specialized SMT system would be required.

2

Statistical Machine Translation and Comparable Corpora

This chapter provides a brief overview of statistical machine translation, discusses textual components and corpora, describes the Moses tool environment for SMT, discusses major aspects of SMT processing, and describes the evaluation of SMT output.

2.1 Overview of SMT

Statistical machine translation is a method of translation in which a translation is created on the basis of statistical models, which are obtained through an analysis of parallel corpus of a language pair [23].

A parallel text (bilingual or multilingual) is aligned, at least at the sentence level, in at least two languages. Commonly, one of the texts is the original one and the others are its translations, although sometimes it can no longer be determined which text is the original, because they were all developed in parallel and mutually adjusted [23].

Generally, parallel texts can also be understood as different texts with similar topics or forms, in which the alignment of sentences is not possible. However, a comparison of them may prove useful for a linguist, terminologist or translator. Even if initially the entries are purely translations of a text in another language, they are supplemented and developed independently within each language community and usually function independently of each other. Such data is called a comparable corpus [5].

Comparable corpora are often considered a solution to the problem of an insufficient amount of parallel data. In fact, comparable corpora are perceived as the reference source when using multilingual natural language processing, such as statistical machine translation [13].

Large sets of parallel texts are called parallel corpora. Alignment of parallel corpora at the sentence level is necessary for conducting research in many fields of linguistics. However, it is vital to remember that the alignment of specific segments of the translation is not always easy, because those parts of the original text might have been separated, joined, removed, moved or replaced with a completely new element by the translator [13].

Probability theory constitutes the foundation of statistical translation. During the translation of a document, the probability p(e|f) that a string f that appears in the source language (e.g., Polish) is the translation of the string e in another language (e.g., English) is defined. Using Bayes' theorem for p(e|f), it can be said that it is proportional to p(f|e)p(e), where the model of translation p(f|e) is the probability that the foreign text is a translation of the source text, and the language model p(e) is the probability of occurrence of a target text. Therefore, the initial problem is split into two sub-problems. Hence, finding the best translation $\tilde{e}$ is simply finding the one with the highest probability:

$$\tilde{\tilde{e}} = {}^{\arg max}_{e \in e*.} \, p(e \mid f) \sim {}^{\arg max}_{e \in e*.} \, p(f \mid e)p(e) \tag{1}$$

Statistical models are determined using specific tools with training conducted on parallel corpora. The resulting n-gram models predict the next element of a sequence. The creation of an n-gram model starts with counting the occurrences of a sequence of a determined length n in the existing language resource. Usually, entire texts are analyzed, and all 1's (1-grams, unigrams), 2's (2-grams, bigrams), and 3's (3-grams, trigrams), etc., are counted. A significant amount of language data is required in order to obtain 4-grams of words, which is especially difficult to achieve in the case of the Polish language [9].

After analyzing a sufficiently large amount of text, the number of occurrences is changed into probability through standardization. Such action makes it possible to predict the next element on the basis of the sequence of n observed so far. In accordance with [208] for the Polish language, analysis of texts containing 300 million words provides a good 1-gram model and a satisfactory 2-gram model.[5] To create a reliable 3-gram model, much larger resources are required. A huge amount of analyzed text increases the quality of the model. However, there are also methods that allow enhancement of n-gram models without additional data, due to smoothing of the gathered statistical information [9].

It must be noted that there are problems in balancing weights between infrequent and frequent n-grams. Items not seen in the training data will be given a probability of 0.0 without smoothing. In practice, it is

---

[5] https://pl.wikipedia.org/wiki/N-gram

also necessary to smooth the probability distributions by assigning non-zero probabilities to unseen n-grams. This approach is motivated by the fact that models derived from only the n-gram counts would have big problems if confronted with any n-grams that have not been seen before [171].

The main advantages of n-grams are simplicity and scalability. By changing n, it is possible to achieve both: Models that do not require a lot of training data, but also do not provide a high predictive accuracy, as well as models that require a large amount of data, but offer greater prediction potential [9].

Some of the most commonly cited advantages of statistical translation are:

- Better use of resources: There are an enormous number of digital texts in a natural language ready for processing by the translation systems.
- Creation of systems based on rules requires the manual development of language rules, which involves a large amount of human resources and, therefore, can be expensive. In addition, such systems cannot be easily adjusted to support languages different from the ones taken into account while creating the system, or for application to a different domain.
- Statistical translation systems are based on texts translated by a human. Therefore, such translation sounds more natural than a text translated using pre-programmed rules [85].

2.2 Textual components and corpora

This section discusses the basic components of texts and the nature of corpora.

2.2.1 Words

Naturally, the fundamental unit of written language (in European languages) that native speakers can identify is a word, which is also used in SMT (In spoken language, basic units are syllables and phrases). A good example is the word "house" which means a dwelling that serves as living quarters for one or more families. In terms of structure, it may translate to a structure that has a roof and walls and stands more or less permanently in one place. While the word (house) may be used in different contexts that surround the language unit and help to determine its interpretation, the word more often than not conveys the same meaning [85].

That aside, there are intricate language units, known as syllables, used in a word (house). More precisely, "s" may not add any weight to the word, even though the word can be understood by native English speakers. All in all, words are separated to form a coherent meaning by the way they appear independently. However, the same cannot be said about a foreign language. Compellingly, when listening to a person of a different native language speak—perhaps a language some person is not used to—people tend to get lost in the syllabic formation of the words. The words often sound different from what people are used to. A case in point is the Polish language, which may sound separated in speech but appear without spaces in writing (e.g., "nie rób"—"don't do"—is spoken the same way as "nierób"—"lazy person"). Homework, for instance, is an English word that stands for work students do at home. Although it appears as two words, it is actually one [85].

Another aspect that complicates consideration of the concept of words as immeasurably small units of criticality is casual representations. A good example is the following sentence: "She's gone off the deep end." The words in the statement "off the deep end" all have diverse meanings that are not related to the intended meaning, which can be confusing. Note that this issue of non-compositional declarations is an obstacle to understanding. While it is acknowledged that people can frequently decipher word-by-word, it is clearly infeasible for such cases [85].

2.2.2 Sentences

Sentences are strings of words that satisfy the grammatical rules of a language. Below is an example:

Jane purchased the house

The focus of this sentence is the verb "purchased." It has a purchaser and an article to be purchased: The subject, "Jane," and the object, "the house." Different verbs refer to various numbers of objects. Some do not require any object whatsoever (known as intransitive verbs), a case in point being "Jane swims." Some require two objects, for example, "Jane gave Joe the book." What and how many objects a verb requires is known as the valency of the verb. Objects needed by a verb are known as arguments. Adjuncts add information to a sentence. The sentence "Jane purchased the house" may be expanded by a prepositional phrase, for example, "from Jim" or "without a second thought," or verb modifiers, for example, "yesterday" and "inexpensively." It is often not simple to differentiate between adjuncts and arguments. Also, distinct meanings of a verb may have diverse valency [85].

Arguments and adjuncts may be complex. Noun phrases (e.g., "the house") may be modified by descriptive words (e.g., "the beautiful house") or prepositional representations (e.g., "the house in the wealthy neighborhood").

The English language calls for a critical understanding of a few basic tenets. Some important tenets are as follows:

FRAGMENT—Sentence fragments are among the most common grammar mistakes that one can find. Essentially, a fragment is characterized by a broken structure and meaning, and it cannot stand independently. Lacking a proper subject and verb association, sentence fragments fail to meet the standards of a complete sentence and meaningful statement. An example of a sentence fragment follows below:

Fox News, during the last American general elections

Whereas the sentence above only provides a subject (Fox) and setting and time (the last general elections), it fails to meet the requirements that were discussed earlier. In this case, it is necessary to introduce a link between the two thoughts. The final sentence may be:

Fox News was ranked among the top networks during the last American general elections

The sentence above is complete, having a subject, verb and place.

Be that as it may be, stylistic fragments can be used in creative writing, which is less restrictive than formal writing [85].

RUN-ON—Also known as fused sentences, run-on sentences are typified by putting two complete sentences together without a break or punctuation. An example of a run-on sentence is:

My favorite movies are documentaries they are very educational

The sentence above should be broken into two by using a comma, a coordinating conjunction or a period. For instance, the sentence could read:

My favorite movies are documentaries, for they are very educational

My favorite movies are documentaries; they are very educational

My favorite movies are documentaries. They are very educational

COMMA SPLICE—As with the above grammatical errors, a comma splice occurs when two independent clauses are treated as one sentence, as opposed to breaking them into two. Correcting comma splices can be done by using a comma, semicolon, period or conjunction.

Example: *My friends come to my place every Saturday morning, we then go swimming together.*

Correction: *My friends come to my place every Saturday morning; we then go swimming together. My friends come to my place every Saturday morning. We then go swimming together.*

SEMICOLON—This punctuation mark is used to clarify meaning by separating words into similar sentences, clauses or phrases. Importantly, they provide breaks in rather long sentences. A period or comma can replace a semicolon.

Example: ***James is a music producer; writing songs is one of his fortes.***

The semicolon here has been used to show a symbiotic relationship between independent clauses.

COMMA RULES—Arguably, these are the most often-used punctuation marks, with the exception of periods, of course. That being the case, there are a few rule of thumbs applied in their usage. They are:

Rule 1. To break up a list of items
Example: ***Elizabeth is known for her wits, skills and personality.***

Rule 2. To separate two adjectives (especially when they are identical)
Example: ***Africans are warm, friendly people.***

Rule 3. To separate two independent clauses
Example: ***Hannah ran to the studio, and talked to the photographer.***

In this case, the comma is employed to give a rest to the first clause, thus avoiding a run-on or comma splice.

Rule 4. Used in combination with introductory words, such as "well," "yes," "in fact," "surprisingly," etc.
Example: ***Surprisingly, Joshua did not even attend his own graduation.***

Rule 5. Used in combination with connectors or statements that suspend the flow of the sentence
Example: ***South Africa is a lovely country to live in; however, the high level of insecurity is a great concern.***

Rule 6. Used to set apart a name or title of a person
Example: ***Christine, will you marry me?***
Example: ***Sir, may I submit my assignment tomorrow?***

Rule 7. Used to separate dates
Example: ***The last time we spoke was on February, 10, 2013.***

Rule 8. Used to introduce or interject a direct quote
Example: ***She said, "Please don't call me ever again."***

These developments can be again seen as additional aspects. Also, they can also be extended: Modifiers by intensifiers ("brilliant" becomes "astoundingly brilliant"), noun phrases by additional prepositional phrases ("the house in the rich neighborhood over the stream"), etc. The ability to make such subtle nuances about phrases, which can be connected by additional phrases, is a striking idiosyncrasy of language called recursion. Note that this can also be applied to sentences. This exceptional sentence may be extended by modifiers of Jane, who, starting late, won the lottery and acquired the house that was just put on the market [85].

This provides an opportunity to refine the considered sentence. A sentence may involve one or more modifiers, each of which embodies a verb with discords and subordinates. Stipulations may be partners and controversies themselves, as in: "I proposed to set out for some swimming." The recursive refinement of sentences with subtle improvements raises a huge number of issues for the examination of text. One especially grievous issue for specific sentences is structural obscurity. Consider the three sentences:

- *Joe eats steak with a knife.*
- *Jim eats steak with ketchup.*
- *Jane watches the man with the telescope.*

Each of the three sentences end in a helping prepositional phrase. In the first sentence, the helper modifies the verb (the consumption happens with a knife). In the second sentence, the subordinate clause modifies the object (the steak has ketchup on it). Be that as it may, what about the third sentence? Does Jane utilize a telescope to watch the man, or does the man she watches have a telescope? Structural uncertainty is presented not only by helpers, for example, the prepositional expression issue mentioned previously. Connectives likewise add ambiguity, as their extension is not generally clear. Consider the sentence: Jim washes the dishes and watches television with Jane. Is Jane assisting with the dishes, or would she say that she is simply joining Jim for TV [85]?

The motivation behind why structural ambiguity is such a difficult issue for programmed common dialect preparation frameworks is that the uncertainty is usually determined semantically. People are not confused, reading in context. In addition, speakers and authors use ambiguity when it is not clear what the right intention is. However, what is clear to a human is not evident to a machine. In what manner ought your desktop computer to realize that "steak and knife" do not make a delicious dinner [85]?

2.2.3 *Corpora*

The multilingual nature of the world makes translation a crucial requirement today. Parallel dictionaries constructed by humans are a widely-available resource, but they are limited and do not provide enough coverage for good quality translation purposes, due to out-of-vocabulary words and neologisms. This motivates the use of statistical translation systems, which are dependent on the quantity and quality of training data. A corpus is a large collection of texts, stored on a computer. Text collections are called corpora. The term "parallel corpus" is typically used in linguistic circles to refer to texts that are translations of each other. For statistical machine translation, parallel corpora, which are texts paired with a translation to another language, are of particular interest. Texts differ in style and topic, for instance, transcripts of parliamentary speeches versus news reports. Preparing parallel texts for the purpose of statistical machine translation may require crawling the web, extracting the text from formats such as HTML, and performing document and sentence alignment [7].

A noisy parallel corpus contains bilingual sentences that are not perfectly aligned or have poor-quality translations. Nevertheless, mostly bilingual translations of a specific document should be present in it. To exploit a parallel text, some kind of text alignment, which identifies equivalent text segments (approximately, sentences), is a prerequisite for analysis.

In a comparable corpus, the texts are of the same kind and cover the same content, but they are not translations of one another [5]. An example is a corpus of articles about football from English and Polish news monographs. A quasi-comparable corpus is a type of comparable corpus that includes very heterogeneous and non-parallel bilingual documents that may or may not be topic-aligned [13]. Machine translation algorithms for translating between a first language and a second language are often trained using parallel fragments, comprising a first language corpus and a second language corpus, which is an element-for-element translation of the first language corpus. Such training may involve large training sets that may be extracted from large bodies of similar sources, such as databases of news articles written in the first and second languages describing similar events. However, extracted fragments may be comparatively "noisy," with extra elements inserted in each corpus. Extraction techniques may be devised that can differentiate between "bilingual" elements represented in both corpora and "monolingual" elements represented in only one corpus, and for extracting cleaner parallel fragments of bilingual elements. Such techniques may involve conditional probability determinations on one corpus with respect to the other corpus, or joint probability determinations that concurrently evaluate both corpora for bilingual elements [13].

Because of such difficulties, high-quality parallel data is difficult to obtain, especially for less popular languages. Comparable corpora are a solution to the problem of lack of data for the translation systems for under-resourced languages and subject domains. It may be possible to use comparable corpora to directly obtain knowledge for translation purposes. Such data is also a valuable source of information for other cross-lingual, information-dependent tasks. Unfortunately, parallel data is quite rare, especially for the Polish-English language pair. On the other hand, monolingual data for those languages is accessible in far greater quantities [98].

The domain of a text can significantly affect SMT performance. An important example is medical texts. Obtaining and providing medical information in an understandable language is of vital importance to patients and physicians [63–65]. For example, as noted by GGH [66], a traveler in a foreign country may need to communicate their medical history and understand potential medical treatments. In addition, diverse continents (e.g., Europe) and countries (e.g., the U.S.) have many residents and immigrants who speak languages other than the official language of the place in which they require medical treatment.

Karliner [67] discusses how human translators could improve access to health care, as well as improve its quality. However, human translators trained in medical information are often not available to patients or medical professionals [68]. While existing machine translation capabilities are imperfect [68], machine translation has the promise of reducing the cost of medical translation while increasing its availability [69] in the future. The growing technologies of mobile devices hold great promise as platforms for machine translation aids for medical information.

Medical professionals and researchers, as well as patients, have a need to access the wealth of medical information on the Internet [63, 70]. Access to this information has the potential to greatly improve health and well-being. Sharing medical information could accelerate medical research. English is the prevalent language used in the scientific community, including for medical information, as well as on the Internet, where a vast amount of medical information may be found. The experiments on developing medical-related translation systems were also conducted and are described in the Chapter 4.9.2.3.

2.3 Moses tool environment for SMT

Moses is a tool environment for statistical machine translation that enables users to train translation models for any two languages. This implementation of the statistical approach to machine translation is currently the dominant approach in this field of research [8].

To train a system, a collection of translated texts aligned into a parallel corpus is required. The trained model is enough for an efficient search algorithm implemented in Moses to quickly find the most probable sentence translation from a very large number of possible translation candidates [8].

In the training step, Moses processes the parallel data and analyses co-occurrences of words or entire segment phrases in order to determine the translation equivalents between two languages. In the most popular phrase-based machine translation, the correspondences between languages are analyzed solely between continuous word sequences. On the other hand, in hierarchical phrase-based machine translation and syntax-based models, some additional structure is added to the relationships. For example, when using a hierarchical MT system, it is possible to "learn" that the Polish "Sok z X został wypity" corresponds to the English "Juice from X was drunk," where X can be replaced by any Polish-English word pair. It is possible to obtain such extra structure through the linguistic analysis of the parallel data. Additionally, an extension to phrase-based machine translation is implemented in Moses. It is known as a factored translation model, and it enables a phrase-based system to be enriched with linguistic information (e.g., Part of Speech tags) [8].

Nonetheless, the key to creating a high-quality system is usage of the best possible quality data in large quantities. It is possible to find free sources of bilingual data for the Polish-English pair. However, it is essential to choose data that is very similar to the data that the system is supposed to translate. Proper data selection will greatly improve the results of the translation. This is also an advantage of using an the open-source Moses toolkit, because it enables users to tailor systems to specific needs with the potential to get better quality and performance than a general-use translation engine. Moses requires data aligned at least at the sentence level for its training process. However, if data is aligned only at the document level, it can often be adjusted. This monograph addresses how to extract parallel data from comparable corpora and a method to make such a conversion [8].

Moses can be divided into two main components, known as the training pipeline and the decoder. Of course, there is much more to the Moses toolkit. It also includes a variety of tools and utilities. In fact, the training pipeline is, in reality, a chain of collected tools that can process text data and turn it into a translation model. Just the opposite, the decoder is a single tool that translates source documents into another language [8].

It is typically required that data be prepared before it is used in training. Steps like tokenization and casing standardization need to be performed. Usually, heuristics that remove sentence pairs that might be misaligned and that remove sentences that are too long are also used. When the pre-processing of the data is done, the remaining bi-sentences are word-aligned. Typically, because of its popularity and quality, the GIZA++ tool, which is an implementation of a set of statistical models developed at IBM, is used for alignment [185, 186]. Phrase-to-phrase translations must be developed from these word alignment models, or hierarchical rules must be developed. Additionally, statistics of the entire corpus are estimated in order to count the probabilities of phrase translations [11]. More details are discussed in the Chapter 2.4.

Tuning is the last step in the creation of a translation system and can improve its quality. The purpose of tuning is to select the best possible weights between trained statistical models. Most popular tuning algorithms are already implemented in Moses [8]. The batch margin-infused relaxed algorithm (MIRA), developed by Cherry and Foster [198], is an algorithm that efficiently parallelizes SMT system tuning. Batch-MIRA has been implemented in the Moses toolkit.

Lastly, the decoder can be used to find the sentence with the highest score in accordance with the translation model. It can also return a sorted list of translation candidates, together with various types of information about how the candidates were found. Decoding is an enormous and time-consuming search problem—too big to conduct an exact search. Moses provides a few solutions to this problem; the most important are stack-based search, cube-pruning, chart parsing, etc. [8].

Sentence translation probability can by calculated as follows:

$$e_{best} = \text{argmax}_e \prod_{i=1}^{I} \phi(\bar{f}_i \mid \bar{e}_i)\, d(\text{start}_i - \text{end}_{i-1} - 1) p_{LM}(\mathbf{e}) \qquad (2)$$

In accordance with the above equation, several components contribute to overall translation probability. To be more precise, the phrase translation probability is defined as ϕ, the reordering model is denoted by d, and the language model is designated as p_{LM}. The input phrases are denoted as $\bar{f}_i$, the output phrases as $\bar{e}_i$, and their positions as start$_i$ and end$_i$, where i = 1, ..., I. It is also possible to compute partial scores for partial translations during the translation process. In other words, it is not only possible to compute the translation probability for the whole translation only when it is completed, but also to incrementally add its parts while determining the final translation. In such a scenario, scores from the three model components need to be considered each time when adding a phrase translation. A reordering model is consulted in order to keep track of the end position in the input of the previous phrase that was translated (end$_{i-1}$). In addition, a place from which a phrase was picked is known, just as its start position in the input (start$_i$). Having those two values, it is possible to calculate the distortion probability distribution d to get the reordering cost. Lastly, a language model is used to count, for example, the probability p_{LM} (he| <s>) that a sentence started with "he", the probability p_{LM} (does|he) that "he" is followed by "does," and so on. In other words, once new words are added to the end of the sequence, their language model impact is computed.

The phrase-based decoder implemented in Moses uses a beam search algorithm inspired by [207]. In this approach, an output sentence e is generated from left to right in the form of hypotheses. As presented in Figure 1, starting from the initial hypomonograph, the first expansion is the *f* word Maria, which is translated as "Mary." A translated word is marked with the "*" symbol.

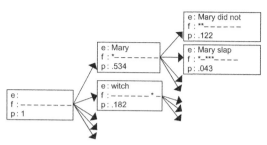

Figure 1. Phrase translation example.[6]

It is possible to generate new hypotheses from previously expanded ones. By doing so, the first two *f* words become marked as translated. By following the back pointers of the hypotheses, it is possible to read partial translations of the sentence.

In the Moses decoding beam search algorithm, a separate stack is created for each *f* word covered in the translation. The first translation hypomonograph is placed on the stack, not having any foreign words covered, but starting with a hypomonograph such that all new ones are generated by committing to phrasal translations that covered previously-unused *f* words. The hypotheses are placed on the stack by the number of *f* words they cover.

2.3.1 *Tuning for quality*

Undoubtedly, the best way to obtain great performance from an SMT system is by using a decent translation phrase table. However, tuning of the decoder parameters is also possible. The statistical cost that is estimated for each translation is a result of the combined costs of four main models. The phrase tables, monolingual language models, reordering models and word penalties must be taken into account [8].

Each of these models helps by supplying the decoder with information on each aspect of a translation. The phrase table guarantees that native phrases and target expressions are precise equivalents of one another in two different languages. The usage of a language model guarantees that the output is fluent language. By using a distortion model, the decoder takes reordering of the input data into consideration. Finally, a word penalty guarantees that the final translation is not excessively long or simply too short [8].

The importance of each of the components can be given proper weights that set their influence level. Influence can be calculated as:

---

[6] http://www.statmt.org/moses/?n=Moses.Background

$$p(e \mid f) = \phi(f \mid e)^{weight_\phi} \times LM^{weight_{LM}} \times D(e, f)^{weight_d} \times W(e)^{weight_\phi} \quad (3)$$

where $\phi(f \mid e)$ is phrase translation, LM is the language model, D is the distortion model, and W is the word penalty.

The Moses decoder uses such weights by passing four parameters: weight-t, weight-l, weight-d, and weight-w. By default, the weights are set to 1, 1, 1 and 0, respectively. Adjusting these weights to the best possible values for each translation system can enhance translation accuracy [8].

2.3.2 Operation sequence model (OSM)

The operation sequence model (OSM) is integrated into Moses [62, 71]. The OSM is an n-gram-based translation and reordering model that can represent aligned bilingual corpora as sequences of operations and that can learn a Markov model over the resultant sequences.

The OSM can perform operations such as the generation of a sequence of source and target words. It can also inspect gaps as explicit target positions for reordering operations, and forward and backward jump operations that do the actual reordering. An n-gram model describes the probability of a sequence of operations, i.e., the probability of an operation depends upon the first n–1 operations. Let $O = o1, \ldots, oN$ be a sequence of operations as hypothesized by the translator to generate a word-aligned bilingual sentence pair <F;E;A> [71]. Given a bilingual sentence pair (F, E) and their alignment (A), the model is defined as:

$$P_{OSM}(F, E, A) = p(o1,...,oN) = ip(o_i \mid o_i - n + 1...o_i - 1) \quad (4)$$

The OSM model addresses a couple of issues with phrase-based SMT and lexicalized reordering models. First of all, it considers target and source information, based on context and crossover phrasal boundaries, and avoids assumptions regarding independencies. It is also focused on minimal translation units, so it does not have the issue of spurious segmentation of phrases. Lastly, the OSM considers much richer condition formation than the lexicalized reordering model, which simply learns the orientation of a phrase while ignoring information on how previous words were reordered and translated. The OSM model conditions translation and reordering decisions on n defined past reordering and translation decisions that can cross over phrasal boundary points [71].

Due to information sparsity, the lexically-determined OSM model may frequently fail in small context sizes. This issue is explored in [80] by learning sequences of operations rather than using generalized representations (for example, POS/Morph or tags/word classes).

2.3.3 *Minimum error rate training tool*

The Minimum error rate training (MERT) technique, introduced by Och, is a tuning algorithm for SMT systems, designed to minimize translation errors by optimizing feature weights and leveraging multiple reference translations [203]. An implementation of MERT is provided in the Moses toolkit, which is widely used [210]. MERT is a form of coordinate descent that attempts to find a local minimum of a translation metric by conducting a line search in different directions [210]. This enables refined tuning of an SMT system.

2.4 Aspects of SMT processing

In the following chapter, technical aspects of statistical machine translation process will be described.

2.4.1 *Tokenization*

Among the most important steps in text translation is tokenization, the dividing of rough text into words. For lingos and languages that use the Latin letters in place, this is basically an issue of dividing a combination of many words. It is a very difficult task to compose frameworks that do not provide spaces between words. This section discusses some of the issues that must be addressed in tokenization.

Sentences are a fundamental unit of text. Determining where sentences end requires recognition of punctuation marks, very often the period ("."). Unfortunately, the period presents a high degree of ambiguity. For example, it can serve as the end of a sentence, part of an abbreviation, an English decimal point, an abbreviation and also the end of a sentence, or be followed by quotation marks [215].

Recognizing abbreviations, which can come from a long list of potential abbreviations, is a related problem, since the period often appears in them. There are also many potential acronyms that can occur in text and must be detected. No universal standard for abbreviations or acronyms exists [215].

Hyphens and dashes present another tokenization challenge. A hyphen may be part of a token (e.g., "forty-five"), or it may not be or (e.g., "Paris-based"). A hyphen is also used for splitting words at the end of a line and within a single word following certain suffixes [216].

A variety of non-lexical expressions, which may include alphanumeric characters and symbols that also serve as punctuation marks, can confuse a tokenizer. Some examples include phone numbers, email addresses, dates, serial numbers, numeric quantities with units, and times [216, 217].

One tokenization issue is words that occur in lowercase or uppercase in a text. For example, the text "house," "House," and "HOUSE" may all occur in a large corpus, in the midst of a sentence, at the beginning of a sentence, or in rare cases, independently. It is the same word, so it is necessary to standardize case, typically by lowercasing or truecasing. Truecasing applies uppercase to names, allowing differentiation between phrases such as "Mr. Fisher" and "a fisher" [85].

Tokenization of modern text must also handle emoticons, which are often used in social media and email messages. There are many different emoticons. They can be complex and may be easily confused with other text. For example, any of the following emoticon symbols may indicate a frown[7]: >:[:(:(: :c :c :< : ⊃ C :< :[:[:{

Much work has gone into the separation of Polish content. Such tasks require a dictionary of possible words and a system to discover, learn and determine them with certainty, usually by consideration of ambiguities between alternative divisions (e.g., favor more successive words). At the same time, there are some tokenization issues for English that do not have simple answers:

- What should be done with the possessive marker in "Joe's," or a contraction, for example, "doesn't"? These are typically separated. For example, the possessive marker "s" is viewed as a word in its own particular right.

- How should hyphenated words be treated, for example, "co-work," "purported," or "high-risk"? There is, by all accounts, little advantage in separating "co-work." However, "high-risk" is truly comprised of two words. When tokenizing content for machine translation, the primary guiding rule is that it is required to reduce content to a grouping of tokens from a small set. It is preferred not to learn diverse translations of "house," contingent upon whether it is trailed by a comma (house) or encompassed by quotes ("house") [85].

2.4.2 Compounding

A few languages, including Polish, support making new words by intensifying existing words. Some of these words—przedszkole, metamaskarada, etc.—have even found their way into the English language. Exasperation is amazingly conducive to the creation of new words. Breaking up compound words may be part of the tokenization stage for a couple of languages [12].

Different words serve different functions. Nouns are used for actual, concrete objects (e.g., house) or abstractions (e.g., opportunity). Words can

---

[7] https://en.wikipedia.org/wiki/List_of_emoticons

be classified as content words or function words. Content words are used for acts, things or properties. Function words describe relationships among content words [85].

Content words are also known as open-class words, since new content words are frequently added to a language as new objects and object types occur. Since this is not true of function words, they are also known as closed-class words. They are added very infrequently (e.g., every few decades) to a language, when at all [85].

Both types of words are challenging for MT in unique ways. Since the huge set of content words change frequently, many large corpora on numerous topics and from different domains are needed for MT. There are only a few function words, and they occur very frequently in extensive corpora. Function words vary significantly from language to language. The most frequent word in English is "the," a function word. Possible translations of "the" in Polish are: "który," "która," "które," "któremu," and "której" [85].

Nouns, verbs and properties are the three types of open-class words. Nouns refer to objects. Verbs refer to actions. On the other hand, properties may be adjectives, which modify nouns, or adverbs, which modify verbs or adjectives [85].

There are several types of closed-class words. As described in [12], they include: Determiners, pronouns, prepositions, coordinating conjunctions, numbers, markers for possession and list items, symbols (e.g., for currency), foreign words, and interjections. They may also include icons, such as the smiley face, ":})" [12].

Natural language processing typically tags each word in a text with its part of speech (POS). To do so, often a large corpus is labeled with POS tags. Then, machine learning techniques use this to train software that tags words in any text with POS [12].

2.4.3 Language models

One essential component of any statistical machine translation system is the language model (LM), which measures how likely it is that a sequence of words would be uttered by a speaker. A language model is a statistical model that is usually built from a large set of monolingual data in a target language. A decoder uses information from the language model to smooth and ensure fluency of the translation. External tools like SRILM [9], IRSTLM [187], KENLM [10], etc., are used in Moses for language modeling [18]. Obviously, it is desirable for a machine translation system to not only produce output words that are true to the original meaning, but also to assemble them into fluent English sentences [8].

The language model typically does much more than enable fluent output. It supports difficult decisions about word order and word translation. For instance, a probabilistic language model p_{LM} should prefer correct word order over incorrect word order:

$$p_{LM}(\text{the house is small}) > p_{LM}(\text{small the is house}) \qquad (5)$$

Formally, a language model is a function that takes an English sentence and returns the probability that an English speaker would produce it. According to the example above, it is more likely that an English speaker would utter the sentence "the house is small" than the sentence "small the is house." Hence, a good language model assigns a higher probability to the first sentence [85].

A language model helps a statistical machine translation system find the right word order. Another area in which language models aid translation is word choice. If a foreign word (such as the Polish "dom") has multiple translations (house, home, etc.), lexical translation probabilities give preference to the more common translation (house). Language models also ensure that the output is fluent. In mathematical terms, LMs add probability to each translated token sequence that indicates how likely it is for that sentence to appear in a real-life text. The LMs used in the Moses decoder are based on n-grams, which use the assumption of Markov to divide the probability of an entire sentence by the probabilities of each word through analysis of the history of preceding words.

In addition, language models are optimized using a perplexity value. Perplexity, developed for information theory, is a performance measurement for a language model. Specifically, it is the reciprocal of the average probability that the LM assigns to each word in a data set. Thus, when perplexity is minimized, the probability of the LM's prediction of an unknown test set is maximized [32, 33, 184, 190].

The fundamental challenge that language models handle is sparse data. It is possible that some possible translations were not present in the training data but occur in real life. There are some methods in Moses, such as add-one smoothing, deleted estimation and Good-Turing smoothing, that cope with this problem [184].

Interpolation and back-off are other methods of solving the sparse data problem in n-gram LMs. Interpolation is defined as a combination of various n-gram models with different orders. Back-off is responsible for choosing the highest-order n-gram model for predicted words from its history. It can also restore lower-order n-gram models that have shorter histories. There are many methods that determine the back-off costs and adapt n-gram models. The most popular method is known as Kneser-Ney smoothing. It analyses the diversity of predicted words and takes their histories into account [8].

As previously mentioned, the leading method for building LMs is n-gram modeling. N-gram language models, in simple terms, are based on statistics that determine how likely it is for words to follow each other. In more formal terms, language modeling counts the probability of a string as $W = w_1, w_2, ..., w_n$. The probability $p(W)$ is defined as the probability that W will be returned when picking a sequence of words at random (by reading random text) [8].

The probability $p(W)$ is computed by analyzing large amounts of textual data and estimating how often the words W occur in it. It must be noted that most long text sequences of words will not be found in the data at all. Because of that fact, the computation of $p(W)$ must be divided into smaller steps. Such smaller steps are superior for collecting the statistics required in order to calculate the probability distribution [8].

A related problem is the compounding of words into new, longer words. Compounding appears in many languages, such as German, Dutch, Finnish and English, and is also present in Polish. For example, "lifetime" is a compound of the words "life" and "time." The problem of merging existing words into new ones that often express different thoughts creates even more sparse data problems for morphologies that are inflectional, such as the Polish language. Words created using this method may have never been seen before or may have been created from foreign words. Even using a very large set of training data will not completely solve this problem [85].

When taking phrase-based language models into account, it is possible to split a compounded word into all possible valid words. This can be done because a phrase-based language model stores many-to-many relations [85].

2.4.3.1 Out of vocabulary words

Another problem that must be addressed when using n-gram language models are out-of-vocabulary words (OOV). In computational linguistics and natural language processing, this term denotes words that were not present in a system's dictionary or database during its preparation [9].

By default, when a language model is estimated, all observed vocabulary is used. In some cases, it may be necessary to estimate the language model over a specific, fixed vocabulary. In such a scenario, the n-grams in the corpus that contain an out-of-vocabulary word would be ignored. The n-gram probabilities would be smoothed over all the words in the vocabulary, even if they were not observed.

Nonetheless, in some cases, it is essential to explicitly model the probability of out-of-vocabulary words by introducing a special token (e.g., "<unk>") into the vocabulary. Out-of-vocabulary words in the corpus are replaced with this special token before n-grams counts are determined. Using this approach, it is possible to estimate the transition probabilities of n-grams involving out-of-vocabulary words [172].

Another recently introduced solution to the OOV problem is unsupervised transliteration. In accordance with [80], character-based translation models (transliteration models) were shown to be quite useful in MT when translating OOV words is necessary, for disambiguation and for translating closely-related languages. The solution presented in [80] extracts a transliteration corpus from the parallel data and builds a transliteration model from it. The transliteration model can then be used to translate OOV words or named-entities.

The proposed transliteration-mining model is actually a mixture of two sub-models (a transliteration model and a non-transliteration sub-model). The transliteration model assigns higher probabilities to transliteration pairs compared to the probabilities assigned by a non-transliteration model to the same pairs. If a word pair is defined as (e, f), the transliteration probability for them is defined as:

$$P_{tr}(e,f) = \sum_{a \in Align(e,f)} \prod_{j=1}^{|a|} p(q_j) \tag{6}$$

Align(e, f) is the set of all possible sequences of character alignments, whereas a is one alignment sequence and q_j is a character alignment.

The non-transliteration model deals with the word pairs that have no character relationship. Multiplying source and target character unigram models, it is as follows:

$$P_{ntr}(e,f) = \prod_{i=1}^{|e|} p_E(e_i) \prod_{i=1}^{|f|} p_F(f_i) \tag{7}$$

The transliteration mining model is an interpolation of the transliteration and the non-transliteration models determined by:

$$P(e,f) = (1 - \lambda) P_{tr}(e,f) + \lambda P_{ntr}(e,f) \tag{8}$$

where λ is the prior probability of non-transliteration [80].

2.4.3.2 N-gram smoothing methods

As previously mentioned, the n-gram estimation assigns a probability of zero when an n-gram is not found in the training text. The problem of zero probabilities can be solved by taking some probability mass from discovered n-grams and distributing it to unobserved ones. Such a solution is known as smoothing or discounting [141].

Most smoothing algorithms can be defined in terms of the estimated conditional probability $p(a_z)$ that the n-th word z will occur given that the first n–1 words of an n-gram (the prefix $a\_$) were observed:

$$p(a_z) = (c(a_z) > 0) \; ? \; f(a_z) : bow(a\_) \, p(\_z) \qquad (9)$$

where a_z is an n-gram starting with word a and ending with word z, $\_z$ is the n–1 word suffix of the n-gram, $f(a_z)$ is a distribution function that represents observations of the n-gram observed in training data, $bow(a\_)$ is a back-off weight[8] for the n-gram, and $p(\_z)$ is a low-order distribution used if the n-gram has not been observed in the training data. The distribution function $f(a_z)$ is discounted so that it is less than the maximum likelihood estimate. Lastly, the "?" symbol in the equation is the ternary operator (a.k.a. the conditional operator), which is similar to an if-then-else construct.

In this equation, if the n-gram a_z was found in the training text, the distribution $f(a_z)$ will be used. Usually, $f(a_z)$ is discounted to be lower than the maximum likelihood estimate [174]. Some probability is left for the z words that were unseen in the context a_z. Different smoothing algorithms differ in the methods they use to discount the maximum likelihood estimate to obtain $f(a_z)$ [85].

If the a_z n-gram was not found in the training text, the lower-order distribution $p(\_z)$ is used. On the other hand, if the context was never found, $(c(a\_) = 0)$, the lower-order distribution can be used directly: $(bow(a\_) = 1)$. In other scenarios, a back-off weight (bow) needs to be computed to ensure that the probabilities will be normalized:

$$\sum_z p(a_z) = 1 \qquad (10)$$

If Z is the set of all words in the vocabulary, Z0 is the set of all words with $c(a_z) = 0$, Z1 is the set of all words with $c(a_z) > 0$, and $f(a_z)$ is given, $bow(a\_)$ can be calculated as:

$$\sum_z p(a_z) = 1 \qquad (11)$$

$$\sum_{z1} f(a_z) + \sum_{z0} bow(a\_) p(\_z) = 1 \qquad (12)$$

---

[8] https://cxwangyi.wordpress.com/2010/07/28/backoff-in-n-gram-language-models/

$$bow(a\_) = \frac{(1 - \sum_{z1} f(a_z))}{\sum_{z0} p(\_z)}$$

$$= \frac{(1 - \sum_{z1} f(a_z))}{(1 - \sum_{z1} p(\_z))} \tag{13}$$

$$= \frac{(1 - \sum_{z1} f(a_z))}{(1 - \sum_{z1} f(\_z))}$$

When Moses is used with the SRILM tool [8], one of the most basic smoothing methods is Ney's absolute discounting. This method subtracts a constant value D, which has a value between 0 and 1. Assuming that Z1 is a set of all words z and $c(a_z) > 0$:

$$f(a_z) = \frac{(c(a_z) - D)}{c(a\_)} \tag{14}$$

$$p(a_z) = (c(a_z) > 0) \; ? \; f(a_z) : bow(a\_) \, p(\_z) \tag{15}$$

$$bow(a\_) = \frac{(1 - \sum_{z1} f(a_z))}{(1 - \sum_{z1} f(\_z))} \tag{16}$$

The suggested factor for discounting is equal to:

$$D = \frac{n_1}{(n_1 + 2n_2)} \tag{17}$$

In this equation, n_1 and n_2 are the total numbers of n-grams that have exactly one and two counts, respectively [85].

Kneser-Ney discounting is similar to absolute discounting if subtracting a constant D from the n-gram counts is considered. The main idea is to use the modified probability estimation of lower-order n-grams that is used for the back-off. More specifically, it is taken with the intention to be proportional to the number of unique words that were found in the training text [142].

Chen and Goodman's modification to Kneser-Ney discounting differs by using three discounting constants for each of the n-gram orders, defined as follows:

$$D_1 = 1 - 2Y(\frac{n_2}{n_1}) \tag{18}$$

$$D_2 = 2 - 3Y(\frac{n_3}{n_2}) \tag{19}$$

$$D_3 = 3 - 4Y(\frac{n_4}{n_3}) \tag{20}$$

where the following factor is used to simplify the above equations:

$$Y = \frac{n_1}{n_1 + 2n_2} \tag{21}$$

Another popular smoothing method used in this research is Witten-Bell discounting. The weights given to the lower-order models should be proportional to the probability of observing unseen words in the context $(a\_)$. Its weights are estimated as:

$$Bow(a\_) = \frac{n(a_*)}{(n(a_*) + c(a\_))} \tag{22}$$

where the expression $n(a_*)$ defines the quantity of unique words present in context $(a\_)$ [85].

In accordance with [209] and empirical study conducted within this research, it was determined that the Kneser-Ney discounting method performs best for lower order n-grams, whereas the Witten-Bell discounting provides better result for n-gram order higher than 6. This is also the reason why Witten-Bell works better when considering morphologically-rich languages.

2.4.4 Translation models

This section discusses existing approaches to translation models. Firstly, noisy channel model is described, next, the IBM translation models are discussed, followed by phrase-based models.

2.4.4.1 Noisy channel model

The noisy channel model, developed in Shannon's information theory for error-correcting codes [212], provides a basis for translation models and their combination with language models. This conceptual model considers an original, clean signal being transferred through a medium, or channel, that adds noise. Thus, the noisy channel corrupts the signal that is received after passing through the channel. In a translation model, for example, English text (e.g., a string) might be considered the original signal, and text in a foreign language that results from translation might be considered the received signal. The task of removing the noise in translation can be viewed as constructing English text e from the foreign text f [85].

Probabilities are often associated with the original signal and the addition of noise in the noisy channel model. Similarly, as discussed in Section 2.1, probabilities can be associated with the translation process. For example, the probability p(e|f) represents the probability that the text f that appears in a foreign language (e.g., Polish) is the translation of the English text e [85].

2.4.4.2 *IBM models*

Five models were proposed in the original work on statistical machine translation at IBM, and the so-called Model 6 was proposed in [178]. The progression of the six models can be summarized as:

- Model 1: Lexical translation
- Model 2: Added absolute alignment model
- Model 3: Added fertility model
- Model 4: Added relative alignment model
- Model 5: Fixed deficiency problem
- Model 6: Model 4 combined with a hidden Markov model (HMM) alignment model in log-linear manner [179].

IBM Model 1 was the first to be introduced for machine translation. After more detailed inspection, the model was determined to have many defects. This model appears to be rather weak in reordering and also in adding and dropping words. According to IBM Model 1, the best translation for any kind of input would be an empty string [177].

Summarized models also introduce a model for reordering words in a sentence. In most cases, words that follow each other in one language should have a different order after translation. Unfortunately, IBM Model 1 treats all kinds of reordering as equally probable [177]. Another problem in alignment is fertility, the notion that input words produce a specific number of output words after translation. In most cases, one input word is translated into a single word, but some words will produce multiple words or even get dropped (i.e., produce no words at all). A fertility of words model addresses this aspect of translation [85]. In the sense that adding components increases the complexity of models, the main principles of IBM Model 1 are consistent. We define the models mathematically and then devise an expectation maximization (EM) algorithm for training [85].

IBM Model 2 provides an additional model for alignment that is not present in Model 1. For example, using only IBM Model 1, the translation probabilities for the translations below would be the same:

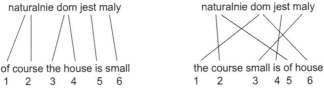

Figure 2. Example of IBM Model 1 alignment.

IBM Model 2 addressed this issue by modeling the translation of a foreign input word in position i to a native language word in position j using an alignment probability distribution defined as:

$$a(i|j, l_e, l_f) \tag{23}$$

In the above equation, the length of the input sentence f is denoted as l_f, and the length of the translated sentence e as l_e. Translation by IBM Model 2 can be presented as a process divided into two steps (lexical translation and alignment).

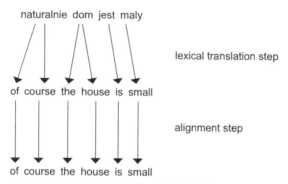

Figure 3. Example of IBM Model 2 alignment.

Assuming that the translation probability is noted as t(e|f) and an alignment probability as $a(i|j, l_e, l_f)$, IBM Model 2 can be defined as:

$$p(e, a \mid f) = \in \prod_{j=1}^{l_e} t(e_j \mid f_{a(j)}) a(a(j) \mid j, l_e, l_f) \tag{24}$$

In this equation, the alignment function a maps each output word position j to a foreign input position a(j) [177].

The fertility of words problem described above is addressed in IBM Model 3. Fertility is modeled using a probability distribution defined as [180]:

$$n(\varphi|f) \tag{25}$$

For each foreign word f, such a distribution indicates how many output words (φ) it usually translates. Such a model deals with the dropping of input words, because it allows the φ be equal to 0. However, there is still an issue when adding words. For example, the English word "do" is often inserted when negating in English. This issue generates a special NULL token that can also have its fertility modeled using a conditional distribution defined as:

$$n(\varphi|NULL) \tag{26}$$

The number of inserted words is dependent on the sentence length. This is why the NULL token insertion is modeled as an additional step, the fertility step. This increases the number of steps in the IBM Model 3 translation process to four steps, defined as:

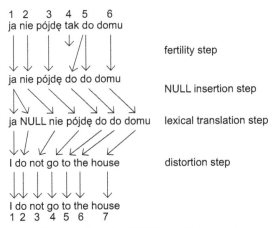

Figure 4. Example of IBM Model 3 alignment.

The last step is called distortion, rather than alignment, because it is possible to produce the same translation with the same alignment in different ways [180].

IBM Model 3 can be mathematically expressed as:

$$P(S|E, A) = \prod_{i=1}^{I} \Phi_i! n(\Phi|e_i)$$

$$* \prod_{j=1}^{J} t(f_j|e_{a_j}) * \prod_{j:a(j)\neq 0}^{J} d(j|a_j, I, J) * \binom{J-\Phi_0}{\Phi_0} p_0^{\Phi_0} p_1^{J-2\Phi_0} \tag{27}$$

where Φ_i represents the fertility of e_i, each source word s is assigned a fertility distribution n, and I and J refer to the absolute lengths of the target and source sentences, respectively [181].

A main objective of IBM Model 4 is to improve word reordering. In Model 4, each word is dependent on the previously-aligned word and on the word classes of the surrounding words. Some words tend to be reordered more than others during translation (e.g., adjective–noun inversion when translating Polish to English). Adjectives often get moved backwards if a noun precedes them. The word classes introduced in Model 4 solve this problem by conditioning the probability distributions of these classes. The result of such a distribution is a lexicalized model.

Model 4 bases the position of a translated source word on the position of the source word that precedes it. The term "cept" is used to denote each source word f_i that is aligned with one or more translated output word. Foreign words with non-zero fertility form cepts. Note that this is not a simplistic mapping, since source words may be inserted, deleted or translated to more than one output word.

Figure 5 provides an example of a set of Polish and English words that have been aligned. This set contains five cepts aligned to English words: ja, nie, idę, do and domu. Table 1 shows each cept, their position (order), the English words to which they are aligned, the English word positions, and the center $\odot$ of the cepts. The center of each cept is the ceiling of the average output word position for the cept.

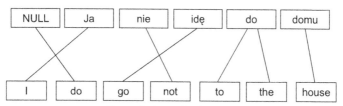

Figure 5. Example alignment.

Table 1. Example cepts and word orders.

Cept π_i	π_1	π_2	π_3	π_4	π_5
Foreign position [i]	1	2	3	4	5
Foreign word $f_{[i]}$	Ja	nie	idę	do	domu
English word $\{e_j\}$	I	go	not	to, the	house
English position $\{j\}$	1	4	3	5, 6	7
Center of cepts $\odot_i$	1	4	3	6	7

Model 4 provides a reordering model that is relative to previously-translated words, based on cepts (aligned words). Mathematically, the Model 4 distortion probability distribution can

For the initial word in a cept:

$$d_1 \left(j - \odot_{[i-1]} | A(f_{[i-1]}), B(e_j) \right) \tag{28}$$

For additional words:

$$d_{>1} \left(j - \pi_{i,k-1} | B(e_j) \right) \tag{29}$$

where the functions $A(f)$ and $B(e)$ map words to their word classes, and e_j and $f_{[i-1]}$ are English and foreign words, respectively. A typical way to define $A(f)$

and B(e) is to use POS tags annotation on the text. A uniform distribution is used for a word generated for a translation that has no equivalent in the source language [85].

Table 2 show the relative distortion for our example.

Table 2. Example relative distortion.

j	1	2	3	4	5	6	7
ej	I	Do	not	Go	to	the	house
In cept π_i, k	$\pi_{1,0}$	$\pi_{0,0}$	$\pi_{3,0}$	$\pi_{2,0}$	$\pi_{4,0}$	$\pi_{4,1}$	$\pi_{5,0}$
$\odot_{i-1}$	0	–	4	1	3	–	6
$j - \odot_{i-1}$	+1	–	–1	+3	+2	–	+1
distortion	$d_1(+1)$	1	$d_1(-1)$	$d_1(+1)$	$d_1(+2)$	$d_{>1}(+2)$	$d_1(+1)$

It must be noted that both Model 3 and Model 4 ignore if an input position is chosen and if probability mass was reserved for the input positions outside the sentence boundaries. This is the reason that the probabilities of all correct alignments do not sum to unity in these two models [179]. These are called deficient models, because probability mass is wasted.

IBM Model 5 reformulates IBM Model 4 by enhancing its alignment model with more training parameters in order to overcome this model deficiency [182]. During translation in Model 3 and Model 4, there are no heuristics that would prohibit the placement of an output word in an already-taken position. In practice, this is not possible. So, in Model 5, words are only placed in free positions. This is accomplished by tracking the number of free positions and allowing placement only in such places. The distortion model is similar to that of IBM Model 4, but it is based on free positions. If v_j denotes the number of free positions in output, IBM Model 5 distortion probabilities would be defined as [183]:

For the initial word in a cept:

$$d_1(v_j | B(e_j), v_{\odot i-1}, v_{max}) \tag{30}$$

For additional words:

$$d_{>1}(v_j - v_{\pi_{i,k-1}} | B(e_j), v_{max}) \tag{31}$$

An alignment model starts with incomplete data—English and foreign words with no connections between them. Model 5 uses iteration of the EM algorithm to iteratively refine alignment. First, it considers model parameters to have uniform probability. In each iteration, it assigns probabilities to missing data and assigns parameters to complete data [85].

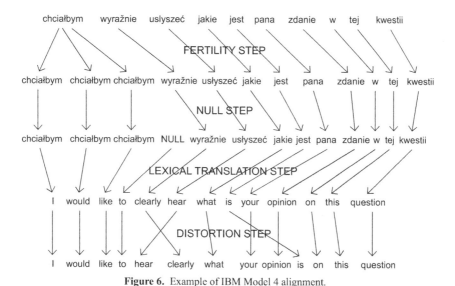

Figure 6. Example of IBM Model 4 alignment.

The alignment models that use first-order dependencies, such as HMM or IBM Models 4 and 5, produce better results than the other alignment methods. The main idea of the HMM is to predict the distance between subsequent source language positions. On the other hand, IBM Model 4 tries to predict the distance between subsequent target language positions. It is expected that the quality of alignment will improve when using both types of such dependencies. This is the reason why the HMM and Model 4 were combined in a log-linear manner in IBM Model 6, as follows [178]:

$$p_6(f,a \mid e) = \frac{p_4(f,a \mid e)^\alpha \cdot p_{HMM}(f,a \mid e)}{\sum_{a',f'} p_4(f',a' \mid e)^\alpha \cdot p_{HMM}(f',a' \mid e)} \qquad (32)$$

where, the interpolation parameter α is used to count the weight of Model 4 relative to the hidden Markov model. A log-linear combination of several models can be defined as $p_k(f,a|e)$ for $k = 1, 2, \ldots, K$ as:

$$p_6(f,a \mid e) = \frac{\prod_{k=1}^{K} p_k(f,a \mid e)^{\alpha_k}}{\sum_{a',f'} \prod_{k=1}^{K} p_k(f',a' \mid e)^{\alpha_k}} \qquad (33)$$

The log-linear combination is used instead of linear combination because the $P_k(f,a|e)$ values are typically different in terms of their orders of magnitude for the HMM and IBM Model 4 [179].

Och and Ney [178] reported that Model 6 produces the best alignments among the IBM models, with lower alignment error rates. This was observed with varying sizes of training corpora. They also noted that the performance of all the alignment models is significantly affected by the training method used [178].

2.4.4.3 Phrase-based models

As discussed in the introduction, using phrases is the most successful approach to MT. By definition, a phrase is a continuous sequence of words. When this approach is used, sentences are divided into word sequences created from phrases. Such phrases require a one-to-one mapping to output phrases to support reordering [85].

In this research, phrase models are estimated from parallel corpora. The GIZA++ tool is used for alignment annotation. Only consistent pairs of phrases are extracted. Probabilistic scores are added based on relative counts or by backing off in accordance with lexical translation probabilities, if necessary [143].

Phrase-based translation models usually do not follow the noisy-channel approach used in word-based models as described in [100], but instead one that facilitates a log-linear framework. Using this approach, it is possible to leverage information such as language models, phrase translation models, lexical translation models and reordering models together with adjusted weights. Such a framework may integrate additional information, such as the number of words created or number of phrase translations used, as well.

Languages differ in their syntactic structure. These differences in word order can be local or global. An example of local reorderings is the swapping of adjectives and nouns in language pairs. Word order changes that span an entire sentence pose a much tougher problem. To generate a good word sequence, a translation system must have very restrictive evidence of how to rearrange the words (e.g., grammar-based translation systems), or have weak evidence in the form of only probabilities (e.g., phrase-based statistical translation systems). Typically, reordering in phrase-based models is determined using a distance-based cost function that discourages reordering in general [175]. Word order changes are usually limited by a maximum number of words. On the other hand, a lexicalized reordering model can distinguish separate reordering behaviors for each analyzed phrase pair. It is also possible to find phrase alignments for sentences by using an expectation maximization algorithm [176] (an iterative method for finding maximum likelihood estimates of parameters in statistical models, in which the model depends on unobserved latent variables).

Statistical phrase-based models assign a score to every possible translation of a phrase. The decoder must find the translation with the highest possible score. Phrase-based models are very well-suited for this job, because the computation of scores is only possible for partial translations. However, this task is still computationally complex (NP-complete complexity). This problem forces restriction of the search space [85].

This problem can be partially solved using a dynamic programming technique that discards hypotheses that might not be part of the best translation. Additionally, it is possible to prune bad translations early by organizing hypotheses in a stack. Limiting the reordering operations might also greatly reduce the search space. During comparison while pruning, it is essential to take the costs of the remaining, untranslated words into account. These words can be counted most efficiently before the decoding phase. Alternatively, stack algorithms, like A* search or the greedy hill-climbing algorithm, first create a rough translation and then optimize it [85].

As mentioned in Chapter 2, the phrase translation model uses the noisy channel model and Bayes rule to reformulate the probability of translating one language sentence f into another language sentence e [206].

During decoding, an input sentence f is segmented into a sequence of I phrases $\bar{f}_1^I$. It is assumed that the probability distribution over all possible segmentations is uniform. Each phrase $\bar{f}_1$ in $\bar{f}_1^I$ must be translated into a phrase $\bar{e}_1$. The phrases e can be reordered. In addition, a phrase-based translation is modeled by a probability distribution defined as: $\phi(\bar{f}_1|\bar{e}_1)$. Due to the use of Bayes rule, the direction of translations is inverted from a modeling standpoint. Reordering of the e output phrases is modeled using a relative distortion probability distribution $d(a_i - b_{i-1})$, assuming that a_i is the start position of the f phrase translated into the i-th e phrase, and b_{i-1} is the end position of the f phrase translated into the (i–1)-th e phrase.

Instead of using only phrase-based models, it is worthwhile to employ a factor-based model that allows additional annotation at the word level, which then may be exploited in various other models. In this monograph, the parts-of-speech tagged English language data sets are used as the basis for the factored phrase model [72]. A factored phrase model extends phrase-based translation models with rich linguistic annotation associated with words that can be leveraged for SMT, instead of processing words as simple tokens alone [85].

Hierarchical phrase-based translation models, which combine the strengths of phrase-based and syntax-based translation, are also popular. They use phrases (segments or blocks of words) as units for translation and use synchronous context-free grammars as rules (syntax-based translation). Hierarchical phrase models allow for rules with gaps. Since these are represented by non-terminals and such rules are best processed with a search algorithm that is similar to syntactic chart parsing, such models fall into the class of tree-based or grammar-based models [85].

For example, we can take the following sentences and rules:

Input: drzwi otwierają się szybko

Rules: drzwi → the door

szybko → quickly

otwierają X1 się → opens X1

X1 X2 → X1 X2

The translation that would be produced is:

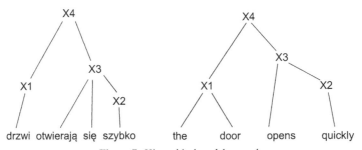

Figure 7. Hierarchical model example.

First, the simple phrase mappings "drzwi" to "the door" and "szybko" to "quickly" are made. This allows for the application of the more complex rule "otwierają X1 się" to "opens X1." Note that, at this point, the non-terminal X, which covers the input span over "szybko," is replaced by a known translation, "quickly." Finally, the glue rule "X1 X2 to X1 X2" combines the two fragments into a complete sentence.

The target syntax model applies linguistic annotation to non-terminals in hierarchical models. This requires running a syntactic parser on the text. The Collins statistical natural language parser [211] is widely used for syntactic parsing. This parser decomposes generation of a parse tree into a sequence of decisions that consider probabilities associated with lexical headwords (considering their parts of speech) at the nonterminal nodes of the parse tree [211]. Bikel [74] describes a number of preprocessing steps for training the Collins parsing model. Detailed descriptions of the parser may be found in [211] and [74].

2.4.5 *Lexicalized reordering*

By default, a phrase-based SMT system uses conditioned reordering, which is based only on movement distance. In addition, some phrases are reordered with higher frequency than others. For instance, adjectives are often switched with preceding nouns. A lexicalized reordering model that is based on conditions determined from actual phrases is more appropriate for MT. The problem of sparse data also arises in such a scenario. For example, phrases may occur only a few times in the training texts. That would make the probability distributions unreliable. It is possible in Moses to choose one out of three possible orientation types for reordering (m—monotone order, s—switch with previous phrase, or d—discontinuous) [85]. Figure 8 illustrates the three orientation types on some sample alignments, designated by dark rectangles [184]:

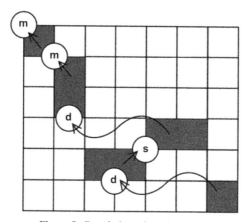

Figure 8. Reordering orientation types.

The Moses reordering model tries to predict the orientation type (m, s, or d) for the phrase pair being translated in accordance with:

$$orientation \ \varepsilon \ \{m, \ s, \ d\}$$

$$p_o \ (orientation | f, \ e)$$

(34)

Such a probability distribution can be extracted from the word alignment. To be more exact, while extracting a phrase pair, its orientation type is also extracted for each specific occurrence. It is possible to detect the orientation type in a word alignment if the top left or top right point of the extracted phrase pair is checked. If the alignment point refers to the top left, it means that the preceding E word is aligned to the preceding F word. In the opposite situation, a preceding E word is aligned to the following F word. Words F and E can be from any two languages [85]. Figure 9 illustrates both situations:

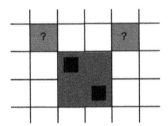

Figure 9. Orientation type detection.

The orientation type can be identified as monotone when the word alignment point in the top left place exists. This is swapped when the word alignment points to the existing top right place. Discontinuous orientation is indicated if there is no top left or top right place that exists. After such an analysis of every phrase pair, appearances of each orientation type are estimated as p_o. The probability distribution p_o is estimated using the maximum likelihood principle, defined as [85]:

$$p_0(orientation) = \frac{count(orientation, e, f)}{\sum_0 count(o, e, f)} \quad (35)$$

Having sparse statistics, it is possible to smooth those counts using an unconditioned maximum likelihood distribution of the probabilities with some factor σ:

$$p_0(orientation) = \frac{\sum_f \sum_e count(orientation, e, f)}{\sum_0 \sum_f \sum_e count(o, e, f)} \quad (36)$$

$$p_0(orientation \mid f, e) = \frac{(\sigma p(orientation) + count(orientation, e, f))}{(\sigma + \sum_0 count(o, e, f))} \quad (37)$$

Numerous orientation types were developed based on this lexicalized reordering. Bidirectional reordering is useful, because some phrases may not score well if they were moved out of order or if subsequent phrases were reordered. It is also possible to condition the probability distribution only to the F phrases (f) or the E phrases (e). In addition, the complexity of the model can be reduced by merging the swap and discontinuous orientation types. This requires making only a binary decision about the phrase order (monotonicity) [85].

Statistics about the orientation collected by analysis of the word alignment matrix are only capable of detecting swaps with previous phrases that contain words exactly aligned on the top right corner, as depicted in Figure 10 [8].

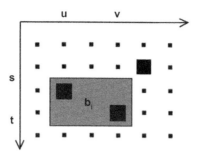

Figure 10. Detection of swaps [115].

Black squares represent word alignments, and gray squares represent blocks identified by phrase-extract [115]. It is assumed that the variables s and t define a word range in a target sentence, and u and v define a word range in a source sentence.

Unfortunately, it is impossible for this approach to detect swaps with phrases that do not contain a word with such an alignment, as shown in Figure 11 [8].

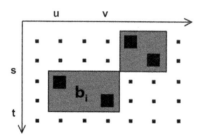

Figure 11. Problems in swap detection [115].

The usage of phrase orientation rather than word orientation in the statistics may solve this problem. Phrase orientation should be used for languages that require a large number of word reordering operations. The phrase-based orientation model was introduced by Tillmann [114]. It uses phrases in both the training and decoding phase. When using the phrase-based orientation model, the second case will be properly detected and counted as a swap. This method was improved in the hierarchical orientation model introduced by Galley and Manning [115]. The hierarchical model has the ability to detect swaps and monotone arrangements between blocks that can be longer than the length limit applied to phrases during the training or decoding step. It can detect the swaps presented in Figure 12 [8].

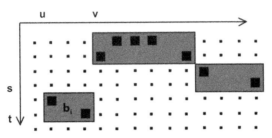

Figure 12. Swap detection in hierarchical model [115].

2.4.5.1 *Word alignment*

One indispensable idea introduced by the IBM models is that of an articulation course of action between a sentence and its translation. In this section, this idea will be further extended, issues with word modification will be pointed out, and measurement of word plan quality will be discussed, a framework for word course of action that uses the IBM models but solves their most glaring issue: Their obstacle to one-to-many alignments [85].

One approach to visualize the task of word arrangement is to use a grid, as shown in Figure 13. Here, alignments between words (for example, between the Polish "pozycję" and the English "position") are depicted by cells in the arrangement matrix.

Word alignments do not need to be one-to-one. Words may have numerous alignment points or none at all. As a case in point, the Polish word "wykorzystał" is aligned to the two English words "took advantage of" The Polish comma is not aligned to any English word, the same as a period [85].

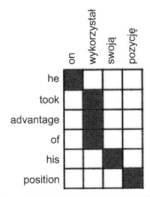

Figure 13. Word alignment example.

On the other hand, it is not generally simple to determine the correct word arrangement. One undeniable issue is that some capacity words have no

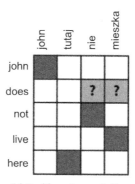

Figure 14. Problematic word alignment.

acceptable alignment in another dialect. See, for example, the illustration in Figure 14.

How would you align the English word "does" in the sentence pair "john does not live here, john tutaj nie mieszka"? Sensible alignments can only be determined by three decisions:

- Since it does not have an acceptable or identical word in one language, it can be concluded that it ought to be unaligned.
- However, the word "does" is, to some degree, joined with "live," since it contains the number of and constrains data for that verb. So, it ought to be aligned with "mieszka," the Polish translation of live.
- A third perspective is that "does" appears in the sentence because the sentence was negated. Without the negative, the English sentence would be "john lives here." So, the word is related to "not," and consequently should be aligned with "nie", the Polish translation of "not."

This provided us an opportunity to research a special word substitution case: Idioms. For instance, consider the sentence pair "jim kicked the bucket, jim kopnął w kalendarz" (see Figure 15).

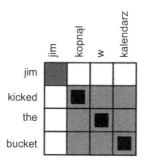

Figure 15. Example of idiom alignment.

This idiom is an everyday expression, synonymous with "died." In this situation, its translation, "kopnął w kalendarz," has the same significance. One may struggle with the fact that a fitting word course of action for this sentence pair alters the verbs "kicked" and "kopnął," the nouns "bucket" and "kalendarz," and the determiners "the" and "w." In any case, shocking content compartment is never a good understanding for "kalendarz," so this is a bit of a stretch. What was presented here is really a phrase that cannot be decomposed further. In the same way, saying "kicked the bucket" is a representation whose essential meaning cannot be contained by decomposing it into its component words [85].

2.4.6 *Domain text adaptation*

Domain adaptation is also very important. Machine translation frameworks are typically constructed for particular domains, for example, film and TV subtitles [101]. A translation model may be prepared from sentences in a parallel corpus that are like the sentences to be translated [102], which may be dictated by dialect model perplexity [103]. Likewise, a specific domain dialect model may be obtained by including only sentences that are like the ones in the target space [104]. Such sentences can be identified utilizing Term Frequency and Inverse Document Frequency (TF-IDF) weighting to test for similarity between sentences as described in [105].

According to Bach et al., sentences may be weighted by how well they match the target space. Instead of discarding a percentage of the training information, a mixture-model methodology weights and consolidates submodels prepared with information from distinct domains [106]. Mixture models might subsequently be utilized to group areas [107]. Such submodels may be joined as parts of the standard log-linear model [108]. To manage different domain-specific translation models, data sentences need to be allocated to the correct domain, which may be done by utilizing a dialect model or data recovery approach [109]. On the off-chance that the classifier accurately judges how well the data falls into a domain, this approach may be utilized to accurately weight the area models [110]. Wu et al. [111] present routines that explore the space of particular monolingual corpora and lexicons. Related to domain adaption is the issue of adjusting machine translation frameworks to diverse territorial dialects that are used [112].

2.4.6.1 *Interpolation*

This research provides a few approaches to gauge the likelihood dispersion for an irregular variable. The most well-known reason for this is that it is easy to obtain a huge corpora with general information and few specific examples, but it is less dependable. So, typically there are two approaches to gauge the likelihood circulation bringing about two functions p_1 and p_2 [9].

Using interpolation, it is possible to consolidate the two likelihood approximations p_1 and p_2 for the same irregular variable X by giving each a fixed weight and including them. If the first is given a weight $0 \leq \lambda \leq 1$, then $1 - \lambda$ is left for the second:

$$p(x) = \lambda\, p_1(x) + (1 - \lambda)\, p_2(x) \tag{38}$$

A typical first application emerges from information aspects that consider distinct conditions. For example, it may be required to predict tomorrow's climate M based on today's climate T and the day D (to consider run-of-the–mill, occasional climate fluctuations) [85].

It is then necessary to gauge the contingent likelihood circulation $p(m|t, d)$ from climate insights. Modeling based on particular days yields little information for use in this estimation. (As a case in point, there may be scarcely any blustery days on August 1st in Los Angeles). Thus, it might make more sense to introduce this distribution with the more robust $p(m|t)$:

$$p_{interpolated}(m|t,\, d) = \lambda\, p(m|t,\, d) + (1 - \lambda)\, p(m|t) \tag{39}$$

2.4.6.2 *Adaptation of parallel corpora*

The adaptation process for parallel corpora is more complicated. When many out-of-domain texts need to be adapted, then the data should be filtered such that the remaining parts of the texts are as similar to the in-domain data as possible. In this research, a method called modified Moore-Lewis [113] filtering was used for this purpose. The method basically trains in-domain and out-of-domain language models, and then removes sentence pairs that receive relatively low scores from the in-domain models. This filtering can be done in two different training phases, either before or after word alignment. Regardless, it should be noted that there may be some benefits of retaining out-of-domain data to improve sentence alignment, but doing so may also be too computationally expensive. It is also possible to specify the percentage of out-of-domain data that will be retained. More details are described in [113].

2.5 Evaluation of SMT quality

To make progress in SMT, the quality of its results must be evaluated. It has been recognized for quite some time that using humans to evaluate SMT approaches is very expensive and time-consuming [25]. As a result, human evaluation cannot keep up with the growing and continual need for SMT evaluation. This led to the recognition that the development of automated SMT evaluation techniques is critical. Evaluation is particularly crucial for translation between languages from different families (i.e., Germanic and Slavic), such as Polish and English [25, 26].

In [26], Reeder compiled an initial list of SMT evaluation metrics. Further research has led to the development of newer metrics. Prominent metrics include: Bilingual Evaluation Understudy (BLEU), the National Institute of Standards and Technology (NIST) metric, Translation Error Rate (TER), the Metric for Evaluation of Translation with Explicit Ordering (METEOR) and the Rank-based Intuitive Bilingual Evaluation Score (RIBES).

2.5.1 Current evaluation metrics

This section provides an overview of existing SMT evaluation metrics, with a focus on BLEU.

2.5.5.1 BLEU overview

BLEU was developed based on a premise similar to that used for speech recognition, described in [27] as: "The closer a machine translation is to a professional human translation, the better it is." So, the BLEU metric is designed to measure how close SMT output is to that of human reference translations. It is important to note that translations, SMT or human, may differ significantly in word usage, word order and phrase length [27].

To address these complexities, BLEU attempts to match variable length phrases between SMT output and reference translations. Weighted-match averages are used to determine the translation score [28].

A number of variations of the BLEU metric exist. However, the basic metric requires determinations of a brevity penalty BP, which is calculated as follows:

$$BP = \begin{cases} 1, c > r \\ e^{(1-r/c)}, c \leq r \end{cases} \tag{40}$$

where r is the sum of the best match lengths of the candidate sentence in the test corpus and c is the total length of the candidate translation corpus [28].

The basic BLEU metric is then determined as shown in [28]:

$$BLEU = BP \exp\left(\sum_{n=0}^{N} w_n \log p_n \right) \tag{41}$$

where w_n are positive weights summing to one, and the n-gram precision p_n is calculated using n-grams with a maximum length of N.

In the BLEU metric, scores are calculated for individual translated segments (generally sentences). Those scores are then averaged over the entire corpus to reach an estimate of the translation's overall quality. The BLEU score is always a number between 0 and 1 [27].

BLEU uses a modified form of precision to compare a candidate translation against multiple reference translations. An over-simplified example of this is:

Test Phrase: "the the the the the the the"

Reference 1 Phrase: "the cat is on the mat"

Reference 2 Phrase: "there is a cat on the mat"

In this example, the precision score is the number of words in the test phrase that are found in the reference phrases (7) divided by the total number of words in the test phrase. This would yield a perfect precision score of 1.

This is a perfect score for a poor translation. BLEU solves this problem with a simple modification: For each word in a test phrase, it uses the minimum of the test phrase word count and the reference word count [27].

If there is more than one reference, BLEU first takes the maximum word count of all references and compares it with the test phrase word count [27]. For the example above:

$$Count \ (\text{``}the\text{''} \ in \ Test) = 7$$

$$Count \ (\text{``}the\text{''} \ in \ Ref \ 1) = 2$$

$$Count \ (\text{``}the\text{''} \ in \ Ref \ 2) = 1$$

BLEU first determines 2 as the maximum matching word count among all references. It then chooses the minimum of that value and the test phrase word count [27]:

$$min(7, 2) = 2$$

BLEU calculates this minimum for each non-repeated word in the test phrase. In this example, it calculates this minimum value just once for word "the". The final score is determined by the sum of the minimum values for each word divided by the total number of words in the test phrase [27]:

$$\text{Final Score} = \left(\frac{2}{7}\right) = 0.2857$$

Another problem with BLEU scoring is that it tends to favor translations of short phrases (on the other hand, SMT usually returns longer translations than a human translator would [197]), due to dividing by the total number of words in the test phrase. For example, consider this translation for above example [27]:

$$Test \ Phrase\text{: ``}the \ cat\text{''} : score = (1 + 1)/2 = 1$$

$$Test \ Phrase\text{: ``}the\text{''} : score = 1/1 = 1$$

BLEU uses a brevity penalty, as previously described, to prevent very short translations. BLEU also uses n-grams. For example, for the test phrase "the cat is here" using n-grams, the following result [27]:

1-gram: "the", "cat", "is", "here"

2-gram: "the cat", "cat is", "is here"

3-gram: "the cat is", "cat is here"

4-gram: "the cat is here"

For the reference phrase "the cat is on the mat", there are, for example, the following 2-grams: "the cat", "cat is", "is on", "on the," and "the mat".

BLEU calculates the score for each of the n-grams. For example, the score calculations for 2-grams follow [27]:

Test 2-grams: "the cat", "cat is", "is here"
Reference 2-grams: "the cat", "cat is", "is on", "on the", "the mat"

It takes:

"the cat": 1

"cat is": 1

"is here": 0

2-grams score = $(1+1+0)/3 = 2/3$

There are several other important features of BLEU. First, word and phrase position within text are not evaluated by this metric. To prevent SMT systems from artificially inflating their scores by overuse of words known with high confidence, each candidate word is constrained by the word count of the corresponding reference translation. A geometric mean of individual sentence scores, with consideration of the brevity penalty, is then calculated for the entire corpus [28]. According to KantanMT[9] and Lavie [144], a score less than 15 means that the MT engine is not providing satisfying quality and a high level of post-editing will be required in order to finalize the output translations and reach publishable quality. A system score greater than 30 is considered a system whose translations should be understandable without problems. Scores over 50 indicate good, fluent translations.

2.5.1.2 *Other SMT metrics*

The NIST metric was designed in order to improve BLEU by rewarding the translation of infrequently-used words. This was intended to further prevent inflation of SMT evaluation scores by focusing on common words and high-confidence translations. As a result, the NIST metric uses heavier weights for

---

[9] http://www.kantanmt.com/

rarer words. The final NIST score is calculated using the arithmetic mean of the n-gram matches between SMT and reference translations. In addition, a smaller brevity penalty is used for smaller variations in phrase lengths. The reliability and quality of the NIST metric has been shown to be superior to the BLEU metric [29].

Translation Error Rate (TER) was designed in order to provide a very intuitive SMT evaluation metric, requiring less data than other techniques while avoiding the labor intensity of human evaluation. It calculates the number of edits required to make a machine translation match exactly to the closest reference translation in fluency and semantics [8, 30].

Calculation of the TER metric is defined in [8]:

$$TER = \frac{E}{w_R} \tag{42}$$

where E represents the minimum number of edits required for an exact match, and the average length of the reference text is given by w_R. Edits may include the deletion of words, word insertion and word substitutions, as well as changes in word or phrase order [8]. It is similar to the Levenshtein distance [85] calculation, but it has the advantage of correctly matching the word order required for correct deletion and re-insertion of misplaced words. Adding an extra editing step that would allow the movement of word sequences from one part of the output to another can solve this. This is something a human post-editor would do with the cut-and-paste function in a text processor, and this is proposed in the TER metric as well [188].

The Metric for Evaluation of Translation with Explicit Ordering (METEOR) is intended to more directly take into account several factors that are indirectly considered in BLEU. Recall (the proportion of matched n-grams to total reference n-grams) is used directly in this metric, to make up for BLEU's inadequate penalty for lack of recall. In addition, METEOR explicitly measures higher order n-grams, explicitly considers word-to-word matches and applies arithmetic averaging for a final score. Arithmetic averaging enables meaningful scores on a sentence level, which can indicate metric quality. METEOR uses the best matches against multiple reference translations [31].

The METEOR method uses a sophisticated and incremental word alignment method that starts by considering exact word-to-word matches, word stem matches and synonym matches. Alternative word order similarities are then evaluated based on those matches. It must be noted that the METEOR metric was not adapted to the Polish language within this research [189]. Such adaptation was done in [219] research, but this was not known at the time of conducting experiments.

METEOR provides a multi-stage algorithm for aligning unigrams. First, all possible unigram alignments are identified. Three approaches to alignment can be used in progressive stages arranged in various orders: Exact unigram matches, unigraphs matches after stemming and matches of synonyms. Within each stage, only unigrams that have not been aligned are evaluated for alignment. Second, METEOR selects the largest set of unigram matches it was able to achieve in its alignment process. If a tie-breaker is needed between sets of matches, METEOR chooses the set with the lowest number of cross-matches between unigrams [31].

After alignment, a METEOR score can be calculated for a translation pair, based on unigram precision and recall. Calculation of precision is similar in the METEOR and NIST metrics as the ratio of the number of aligned unigrams (between translation and reference) to the total number of unigrams in the translation. Recall is calculated at the word level as the ratio of the number of aligned unigrams to the total number of unigrams in the reference translation. To combine the precision and recall scores, METEOR uses harmonic mean. METEOR rewards longer n-gram matches [31]. The METEOR metric is calculated as shown in [31]:

$$\text{METEOR} = \left(\frac{10 \, P \, R}{R + 9 \, P} \right)(1 - P_M) \tag{43}$$

where the unigram recall and precision are given by R and P, respectively. The brevity penalty PM for an alignment is determined by:

$$P_M = 0.5 \left(\frac{C}{M_U} \right) \tag{44}$$

where M_U is the number of matching unigrams, and C is the minimum number of phrases required to match unigrams in the SMT output with those found in the reference translations.

The METEOR score may be calculated for multiple reference translations. The best score is then taken as the score for the machine translation. METEOR scores are aggregated over a test data set to obtain the score for an SMT system [31].

The RIBES metric focuses on word order. It uses rank correlation coefficients based on word order to compare SMT and reference translations. The primary rank correlation coefficients used are Spearman's ρ, which measures the distance of differences in rank, and Kendall's τ, which measures the direction of differences in rank [34].

These rank measures can be normalized to ensure positive values [34]. The normalized Spearman's coefficient is given by:

$$\rho(NSR) = \frac{(\rho+1)}{2} \tag{45}$$

The normalized Kendall's coefficient is determined by:

$$\tau(NKT) = \frac{(\tau+1)}{2} \tag{46}$$

These measures can be combined with precision P and modified to avoid overestimating the correlation of only words that directly correspond in the SMT and reference translations:

NSR Pα and NKT Pα

where α is a parameter in the range $0 < α < 1$ [34].

2.5.1.3 HMEANT metric

Currently correct assessment of translation quality or text similarity is one of the most important tasks in related research areas. Machine translation (MT) evaluation history is quite old but it must be noted that most of the currently applied evaluation approaches and metrics were developed recently. Automatic metrics like BLEU [27], NIST [29], TER [8] and METEOR [31] require only reference translation to be compared with MT output. This resulted in visible acceleration in the MT research field. These metrics, to some extent, correlate with manual human evaluation and, in most cases, are language independent. The main problem in popular MT evaluation metrics is the fact that they analyse only lexical similarity and ignore the semantics of translation [220]. An interesting alternative approach was proposed by Snover [188] and is known as HTER. This metric tries to measure the quality of translation in terms of post-editing actions. It was proved that this method correlates well with human equivalency judgments. However, the HTER metric is not commonly used in MT field because it requires high workload and time.

That is why in 2011 Lo and Wu [220] a family of metrics called MEANT was proposed. It proposes different evaluation approach. The idea is to measure how much an event structure of reference is preserved in translation. It utilizes shallow semantic parsing (MEANT metric) or human annotation (HMEANT) as a high standard.

In this research we implemented HMEANT for a new language (Polish) and we conducted evaluation on the usefulness of the new metric. The practical usage of HMEANT for the Polish language was analysed in accordance with criteria provided by Birch et al. [214]. Reliability in new language was measured as inter-annotator agreement (IAA) for individual stages during the evaluation task. Discriminatory power was estimated as the correlation of HMEANT rankings of our re-speaking results (obtained in manual human evaluation—NER and Reduction, automatic-evaluation—BLEU metric and the HMEANT) that was measured on the level of test sets. The language independence was analysed by collecting the problems of the original method and guidelines to them as described in Bojar and Wu [221] and Birch et al. [214]. Efficiency was studied as the workload cost of annotation task (average time required to evaluate translations) within HMEANT metric. Moreover, we choose unexperienced annotators in order to prove that semantic role labelling (SRL) does not require professional personnel. Being aware that some of the problems existing in HMEANT metric were already outlined [214, 221] and some improvements were already proposed, the decision was made to conduct experiments with native HMEANT form. No changes to the metric itself were done, excepting the annotation interface enhancements that were made.

The first article on MEANT [220] proposed the semi-automatic evaluation approach. It required annotated event structure of two text sequences (reference and translation). The idea was to consider translation correct if it preserved shallow semantic (predicate-argument) structure of reference. Such structure was described in detail in [222]. In simple terms, the evaluation is conducted by probing events in the sentence ("Who did what to whom, when, where, why and how?", etc.). Such structures must be annotated and aligned across two translations. The founders of MEANT reported results of experiments, that utilized human annotation, semantic role labelling and an automatic shallow semantic parser. Their results showed that HMEANT metric correlates with human judgments at the value of 0.43 (Kendall tau, sentence level). This was a very close to the HTER correlation. In contrast BLEU has only 0.20. Also inter-annotator agreement (IAA) was analysed in two stages during the annotation process (role identification and role classification). The IAA ranged from 0.72 to 0.93 (good agreement).

For Czech-English translations MEANT and HMEANT metrics were used by Bojar and Wu [221]. Their experiments were conducted on a human evaluated set containing 40 translations from WMT12 conference. Sets were submitted by 13 different MT systems and each system was evaluated by exactly one annotator (inter-annotator agreement was not examined). HMEANT correlation against human evaluation was equal to 0.28 (much lower than the results of Lo and Wu [220]).

In German-English translation Birch et al. [214] analysed HMEANT in accordance to four criteria, that address the usefulness of the metric in terms of reliability, efficiency, discriminatory power and language independence. The authors conducted experiments on evaluating 3 MT systems using a set of 214 sentences (142 in German and 72 in English). The IAA was divided into annotation and alignment steps. The results showed that the IAA for HMEANT was good enough at the first stages of the annotation but compounding effect of disagreements reduced the effective final IAA to 0.44 for German and to 0.59 for English. The efficiency of HMEANT was stated as reasonably good but it was not compared to other manual metrics.

2.5.1.3.1 Evaluation using HMEANT

The annotation step in HMEANT has two stages: Semantic role labelling (SRL) and alignment. In the SRL phase, annotators are asked to mark all the frames (a predicate and its roles) in reference and translated texts. In order to annotate a frame, it is necessary to mark its predicate (a verb, but not a modal verb) and its arguments (role fillers—linked to that predicate). The role fillers are chosen from the inventory of 11 roles [220] (Table 3). Each role corresponds to a specific question about the entire frame.

Table 3. The role inventory.

Who?	What?	Whom?
Agent	Patient	Benefactive
When?	**Where?**	**Why?**
Temporal	Locative	Purpose
How?		
Manner, Degree, Negation, Modal, Other		

Secondly, the annotators need to align the elements of frames. They must link both actions and roles, and mark them as "Correct" or "Partially Correct" (depending on equivalency in their meaning). In this research we used the original guidelines for the SRL and alignment described in [220].

2.5.1.3.2 HMEANT calculation

Having completed the annotation step, the HMEANT score can be calculated as the F-score from the counts of matches of predicates and corresponding role fillers [220]. Predicates (together with roles) not having correct matches are not taken into account. The HMEANT model is defined as follows:

$\#F_i$—number of correct role fillers for predicate i in machine translation

$\#F_i$ *(Partial)*—number of partially correct role fillers for predicate i in machine translation

$\#MT_i$, $\#REF_i$—total number of role fillers in machine translation or reference for predicate i.

N_{mt}, N_{ref}—total number of predicates in MT or reference

W—weight of the partial match (0.5 in the uniform model)

$$P = \sum_{matched\ i} \frac{\#F_i}{\#MT_i} \tag{47}$$

$$R = \sum_{matched\ i} \frac{\#F_i}{\#REF_i} \tag{48}$$

$$P_{part} = \sum_{matched\ i} \frac{\#F_i(partial)}{\#MT_i} \tag{49}$$

$$R_{part} = \sum_{matched\ i} \frac{\#F_i(partial)}{\#REF_i} \tag{50}$$

$$P_{total} = \frac{P + w * P_{part}}{N_{mt}} \tag{51}$$

$$R_{total} = \frac{R + w * R}{N_{ref}} \tag{52}$$

$$HMEANT = \frac{2 * P_{total} * R_{total}}{P_{total} + R_{total}} \tag{53}$$

In accordance to Lo and Wu [220] and Birch et al. [214] the IAA was studied as well. It is defined as F1-measure, in which one of the annotators is considered to be a gold standard as follows:

$$IAA = \frac{2 * P * R}{P + R} \tag{54}$$

where P is precision (number of labels [roles, predicates or alignments], that were matched between the annotators). Recall (R) is defined as quantity of matched labels divided by total quantity of labels. In accordance to Birch et al. [214], only exact word span matches are considered. The stages of the annotation process described in [214] were adopted as well (role identification, role classification, action identification, role alignment, action alignment). Disagreements were analysed by calculating the IAA for each stage separately.

2.5.2 *Statistical significance test*

In cases where the differences in metrics described above do not deviate much from each other, a statistical significance test can be performed.

The Wilcoxon test [202] (also known as the signed-rank or matched-pairs test) is one of the most popular alternatives to the Student's t-test for dependent samples. It belongs to the group of non-parametric tests. It is used to compare two (and only two) dependent groups that involve two measurement variables.

The Wilcoxon test is used when the assumptions for the Student's t-test for dependent samples are not valid; for this reason, it is considered as an alternative to this test. The Wilcoxon test is also used when variables are measured on an ordinal scale (in the Student's t-test, the variables must be measured on a quantitative scale). The requirement for application of the Wilcoxon test is the potential to rank differences between the first and the second variable (the measurement). On an ordinal scale, it is possible to calculate the difference in levels between two variables; therefore, the test can be used for variables calculated on such a scale. In the case of quantitative scales, this test is used if the distributions of these variables are not close to the normal distribution.

Hypotheses for the Wilcoxon test are formulated as:

$$H_0: F_1 = F_2 \tag{55}$$

$$H_1: F_1 \neq F_2 \tag{56}$$

In this test, as in the case of the t-test, a third variable is used. The third variable specifies the absolute value of the difference between the values of the paired observations. The Wilcoxon test involves ranking differences in measurement for subsequent observations. First, the differences between measurements 1 and 2 are calculated, then the differences are ranked (the results are arranged from lowest to highest), and subsequent ranks are assigned to them. The sum of ranks is then calculated for differences that were negative and for those that were positive (results showing no differences are not significant here). Subsequently, the bigger sum (of negative or positive differences) is chosen, and this result constitutes the result of the Wilcoxon test statistic, if the number of observations does not exceed 25.

For bigger samples, it is possible to use the asymptotic convergence of the test statistic (assuming that H_0 is true) for the normal distribution N(m,s), where:

$$m = \frac{n(n+1)}{4} \tag{57}$$

$$s = \sqrt{\frac{n(n+1)(2n+1)}{24}} \tag{58}$$

The Wilcoxon test is also known as the signed-rank test, because it requires calculation of ranks assigned to different signs (negative and positive differences). As in the case of the Student's t-test for dependent samples, data missing from one measurement eliminates the entire observation from the analysis. Only the observations that have been measured for the first and second time are considered the analysis. The obvious reason is the fact that it is necessary to subtract one result from the other.

3

State of the Art

This chapter will describe current techniques used in statistical machine translation for languages similar to Polish. Recent results in SMT for Polish, as well their improvement in recent years, will also be discussed. In addition, recent methods used for comparable corpora exploration will be described. The native Yalign method, in particular, will be tested and evaluated on a set of selected classifiers.

3.1 Current methods and results in spoken language translation

Competitions accelerate the development of statistical machine translation systems. One of the most popular competitions is the annual meeting organized by the DARPA. However, this competition focuses exclusively on the Arabic-English and Chinese-English language pairs within the news domain. Another annual competition is the International Workshop on Spoken Language Translation (IWSLT).[10] Initially, IWSLT dealt with translating Asian and Arabic languages into English within a limited speech travel domain. Currently, more European languages, as well as the lectures text domain, are part of IWSLT evaluation campaigns [154]. One of the most popular campaigns is the Workshop on Statistical Machine Translation (WMT)[11] competition on European languages, mostly using European Parliament proceedings. Additionally, statistical machine translation research has been conducted by a number of smaller research groups from all around the world.

---

[10] iwslt.org

[11] http://www.statmt.org/wmt14/

Many good quality translation systems already exist for languages that are, to some extent, similar in structure to Polish. When considering such SMT systems for Czech, Slovakian and Russian, it is possible to refer, for example, to a recent Russian-English translation system for news texts that scored almost 32 points in BLEU score [191]. In addition, systems that translated medical texts from Czech to English and from English to Czech, scoring around 37 and 23 BLEU points, respectively, were introduced in [75]. The authors of [193] report BLEU scores of 34.7, 33.7, and 34.3 for Polish, Czech and Russian, respectively, in the movie subtitles domain. The authors of [166] state that they obtained a BLEU score of 14.58 when translating news from English to Czech. It must be noted that those results are not directly comparable with results presented in this monograph, because they were conducted on different corpora.

The development of SMT systems for Polish speech has progressed rapidly in recent years. The tools used for mainstream languages were not adapted for the Polish language, as the training and tuning settings for most popular languages are not suited for Polish. In this research, the trained systems are compared not only to the baseline systems, but also to those of the IWSLT Evaluation Campaign, as a reference point [154]. The data used in the IWSLT are multiple-subject talks of low quality, with many editorial and alignment errors. More information about this domain can be found in Section 4.7. The progress made on PL translation systems during 2012–2014, as well as their quality in comparison with other languages, is shown in Figures 16 and 17. Figure 16 depicts the state of Polish translation systems in comparison to those of other languages in the evaluation campaign (light gray), and the progress made for these systems between 2012 and 2013 (dark

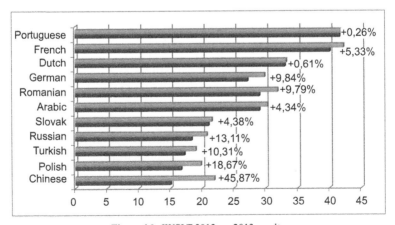

Figure 16. IWSLT 2012 vs. 2013 results.

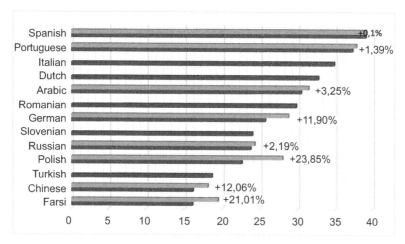

Figure 17. IWSLT 2013 vs. 2014 results.

gray). The SMT system for Polish was one of the worst (mostly because the baseline is rather low); however, through this research, the progress that was made is similar to that of more popular languages. SMT system progress is described by the percentage value beside the bars.

Figure 17 provides similar statistics, but it shows significant progress between 2013 and 2014. In fact, this research resulted in the largest percentage progress during the IWSLT 2014 evaluation campaign. In this campaign, the baseline Polish SMT system advanced from almost last place, and the progress bar indicates that the system could be compared to systems for mainstream languages such as German. It must be noted that the progressive test sets for each IWSLT campaign are different, that is why baseline system scores are different in each year.

3.2 Recent methods in comparable corpora exploration

Many attempts have been made to create comparable corpora (especially for Wikipedia), but none of them addressed the Polish language. Two main approaches for building comparable corpora can be distinguished. Perhaps the most common approach is based on the retrieval of cross-lingual information. In the second approach, source documents must be translated using a machine translation system. The documents translated in that process are then compared with documents written in the target language in order to find the most similar document pairs.

Previous attempts at automatically comparing sentences for parallel corpora preparation were based on sentence lengths, together with vocabulary alignment [41]. Brown's method [43] was based on measuring sentence length by the number of words. Gale and Church [42] measured the number of characters in sentences. Other researchers continued exploring various methods of combining sentence length statistics with alignment of vocabularies [23, 24]. Such methods lead to the creation of noisy parallel corpora, at best.

An interesting idea for mining parallel data from Wikipedia was described in [2]. The authors propose two separate approaches. The first idea is to use an online machine translation system to translate Dutch Wikipedia pages into English, and they try to compare the original English pages with the translated pages. The idea, although interesting, seems computationally infeasible, and it presents a chicken-or-egg problem. Their second approach uses a dictionary generated from Wikipedia titles and hyperlinks shared between documents. Unfortunately, the second method was reported to return numerous, noisy sentence pairs. The second method was improved in [1] by additional restrictions on the length of the correspondence between chunks of text and by introducing an additional similarity measure. They prove that the technique's precision (understood as the number of correct translations pairs divided by the total number of candidates) is about 21%, and in the improved method [4], precision increased to about 43% [2].

Yasuda and Sumita [3] proposed a MT-bootstrapping framework based on statistics that generate a sentence-aligned corpus. Sentence alignment is achieved using a bilingual lexicon that is automatically updated by the aligned sentences. Their solution uses a corpus that has already been aligned for initial training. They showed that 10% of Japanese Wikipedia sentences have an English equivalent.

Interwiki links were leveraged by the approach of Tyers and Pienaar in [4]. Based on the Wikipedia link structure, a bilingual dictionary was extracted. In their research, they measured the average mismatch between linked Wikipedia pages for different languages. They found that their precision is about 69–92%, depending on the language.

In [7], the authors attempt to advance the state of the art in parallel data mining by modeling document-level alignment using the observation that parallel sentences can most likely be found in close proximity. They also use annotation available on Wikipedia and an automatically-induced lexicon model. The authors report recall and precision of 90% and 80%, respectively.

The author of [5] introduces an automatic alignment method, for parallel text fragments found in comparable corpus, that uses a textual entailment technique and a phrase-based SMT system. The author states that significant improvements in SMT quality were obtained (BLEU increased by 1.73) by using this obtained data.

Another approach for exploring Wikipedia was recently described in [6] by Plamada and Volk. Their solution differs from the previously-described methods in which the parallel data was restricted by the monotonicity constraint of the alignment algorithm used for matching candidate sentences. Their algorithm ignores the position of a candidate in the text and, instead, ranks candidates by means of customized metrics that combine different similarity criteria. In addition, the authors limit the mining process to a specific domain and analyze the semantic equivalency of extracted pairs. The mining precision in their work is 39% for parallel sentences and 26% for noisy-parallel sentences, with the remaining sentences misaligned. They also report an improvement of 0.5 points in the BLEU metric for out-of-domain data, and almost no improvement for in-domain data.

The authors in [98] propose obtaining only titles and some meta-information, such as publication date and time for each document, instead of its full contents, to reduce the cost of building the comparable corpora. The cosine similarity of the titles' term frequency vectors were used to match titles and the contents of matched pairs.

In the research described in [99], the authors introduce a document similarity measure that is based on events. To count the values of this metric, they model documents as sets of events. These events are temporal and geographical expressions found in the documents. Target documents are ranked based on temporal and geographical hierarchies. The authors also suggest an automatic technique for building a comparable corpus from the web using news web pages, Wikipedia and Twitter in [100]. They extract entities (using time interval filtering), URLs of web pages and document lengths as features for classifying and gathering the comparable data.

3.2.1 Native Yalign method

In the present research, inspiration was taken from solutions implemented in the Yalign tool, which was modified for better performance and quality. Yalign's native solution is far from perfect but, after the improvements that were made during this research, supplied SMT systems with bi-sentences of good quality in a reasonable amount of time. In addition, two other mining methods were investigated, one that employs bi-lingual analogies detection and another that applies adaptation and a pipeline of text processing tools.

Yalign was designed to automate the parallel text mining process by finding sentences that are close translation matches from comparable corpora. This presents opportunities for harvesting parallel corpora from sources like translated documents and the web. In addition, Yalign is not limited to a particular language pair. However, alignment models for the two selected languages must first be created. It should be noted that the standard Yalign implementation allows the building of only bilingual corpora.

The Yalign tool was implemented using a sentence similarity metric that produces a rough estimate (a number between 0 and 1) of how likely it is for two sentences to be a translation of each other. It also uses a sequence aligner, which produces an alignment that maximizes the sum of the individual (per sentence pair) similarities between two documents. Yalign's initial algorithm is actually a wrapper around a standard sequence alignment algorithm [19].

For sequence alignment, Yalign uses a variation of the Needleman-Wunsch algorithm [20] in order to find an optimal alignment between the sentences in two selected documents. The algorithm has a polynomial time worst-case complexity, and it produces an optimal alignment. Unfortunately, it cannot handle alignments that cross each other or alignments from two sentences into a single one [20].

Since sentence similarity is a computationally-expensive operation, the variation of the Needleman-Wunsch algorithm that was implemented uses an A* approach in order to explore the search space, instead of using the classical dynamic programming method that would require N * M calls to the sentence similarity matrix.

After alignment, only sentences that have a high probability of being translations of each other are included in the final alignment. The result is filtered so as to deliver high-quality alignments. To do this, a threshold value is used. If the sentence similarity metric is low enough, the pair is excluded.

For the sentence similarity metric, the algorithm uses a statistical classifier's likelihood output and normalizes it to the 0–1 range.

The classifier must be trained to determine whether or not sentence pairs are translations of each other. A Support Vector Machine (SVM) classifier was used in the Yalign project. Besides being an excellent classifier, an SVM can provide a distance to the separation hyperplane during classification, and this distance can be easily modified using a Sigmoid function to return a likelihood between 0 and 1 [21]. The use of a classifier means that the quality of the alignment depends not only on the input but also on the quality of the trained classifier.

The quality of alignments is defined by a tradeoff between precision and recall. Yalign has two configurable variables:

- Threshold: The confidence threshold to accept an alignment as "good." A lower value means more precision and less recall. The "confidence" is a probability estimated from a support vector machine, classifying a portion of text as "is a translation" or "is not a translation" of a portion of source text.

- Penalty: Controls the amount of "skipping ahead" allowed in the alignment. Say you are aligning subtitles, in which there are few or no extra paragraphs and the alignment should be more or less one-to-one; then, the penalty should be high. If you are aligning things that are moderately good translations of each other, in which there are some extra paragraphs for each language, then the penalty should be lower [19].

Both of these parameters are selected automatically during training, but they can be adjusted if necessary [19].

Unfortunately, the Yalign tool is not computationally feasible for large-scale parallel data mining. The standard implementation accepts plain text or web links, which need to be accepted as input, and the classifier is loaded into memory for each pair alignment. In addition, the Yalign software is single-threaded [19].

3.2.2 *A* algorithm for alignment*

For the alignment of phrase sequences, calculation of similarity between a pair of phrases can take a significant amount of time using only a single CPU. For this reason, the Yalign tool tries to calculate the best alignment without calculating all the elements of the M matrix. An A* search algorithm is used for this purpose. It is presented in Figure 18 [155].

Instead of dealing with a maximization problem, it is more reasonable to use the minimization equivalent. In such a scenario, an assumption is made that a similarity score of 0 is given to two exact translations, and a score of 1 is given for two completely different translations. In such a problem, the gap penalty is positively added, and the problem becomes a minimization path problem. The principle of this technique is to make a prevision (heuristic), at each point of the path calculation so that maximization can be faster. In the worst case, the number of steps will be equal to that of the Needleman-Wunsch (NW) algorithm [194], but this would happen very rarely (if it was necessary to calculate entire alignment matrix). The objective is to avoid calculation of all matrix elements, but instead only calculate a part of them [155].

Similar to the NW, the A* algorithm starts from the top-left element, goes through all elements that are nearest-neighbors, calculates the lengths of these paths, and uses a heuristic prevision of the best remaining path to the end (bottom-right element). In such a solution, all the path possibilities are left open, but in each iteration, computation is only performed on the path that has the minimal total length (past length + remaining length). The algorithm ends when it reaches the $M_{n,m}$ element. As in the NW implementation, the final step is to backtrack on the best path to the initial position [155].

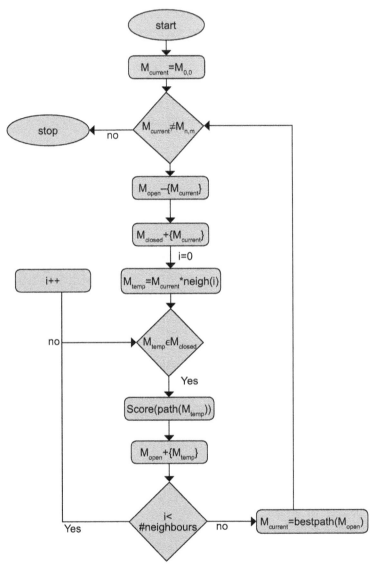

Figure 18. M matrix calculation with A* algorithm.

M_{open} represents the set of all open paths that are represented by their final point. M_{closed} represents the set of all M matrix elements that are no longer the final point of a path, but rather are in the middle of longer paths. $M_{current}$.neigh(i) is the i-th neighbor of $M_{current}$. The minimal total length of the path that ends in M_{temp} is calculated by score(path(M_{temp})). Consideration must also be given to the heuristic for the remaining distance calculation. The bestpath(M_{open})

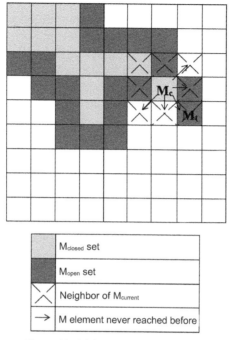

M$_{closed}$ set	
M$_{open}$ set	
Neighbor of M$_{current}$	
→	M element never reached before

Figure 19. A* Search algorithm example.

function gives back the M element that corresponds to the ending point of the minimal length path between all possible paths defined by score(path(...)) in the Mopen set. An example of this algorithm is presented in Figure 19.

3.2.3 Needleman-Wunsch algorithm

The objective of this algorithm is to align two sequences of elements (letters, words, phrases, etc.). The first step consists of defining the similarity between two elements. This is defined by the similarity matrix S, an N × M matrix, where N is the number of elements in the first sequence and M is the number of elements in the second sequence. The algorithm originated in the field of bioinformatics for RNA and DNA comparison. However, it can be adapted for text comparison. The algorithm associates a real number with each pair of elements in the matrix. The higher the number, the more similar the two elements are. For example, imagine that there is the similarity matrix S (phrase-polish, phrase-english) = number between 0 and 1. A = 0 for two phrases means that they have nothing in common; 1 means that those two phrases are the exact translations of each other. The similarity matrix definition is fundamental to the results of the algorithm [20].

The second step is the definition of the gap penalty. This is necessary in the case in which one element of a sequence must be associated with a gap in the other sequence; however, such a step will incur a penalty (p) [20]. In the algorithm, the occurrence of a large gap is considered to be more likely the result of one large deletion than from the result of multiple, single deletions. While an alignment match in an element is given +1 points and a mismatch is given −1 points, each gap between elements is assigned 0 points.

For example, let A_i be the elements of the first sequence, and B_i the elements of the second sequence.

Sequence 1:

A_1, A_2, gap, A_3, A_4

Sequence 2:

B_1, gap, B_2, B_3, B_4

Since two gaps were introduced into the alignment in these sequences and each incurs a gap penalty p, the total gap penalty is 2p (counted as 0 for 2 elements in the alignment).

The calculation of the M matrix (containing alignments of phrases) is performed starting from the $M_{(0,0)}$ element, which is, by definition, equal to 0. The first row of the matrix is defined at the start as:

$$M_{i,0} = -i * p \tag{59}$$

and the first column of the matrix as:

$$M_{0,j} = -j * p \tag{60}$$

After the first row and columns are initialized, the algorithm iterates through the other elements of the M matrix, starting from the upper-left side to the bottom-right side, making the calculation:

$$M_{i,j} = max(M_{i-1,j-1} + S(A_i, B_j), M_{i-1,j} - p, M_{i,j-1} - p \tag{61}$$

Each step of this calculation is shown below:

$M_{up-left}$	M_{up}
M_{left}	$max(M_{up-left} + S(A_{current}, B_{current}), M_{up} - p, M_{left} - p$

If one wanted to calculate the (i,j) element of the M matrix, having already calculated $M_{up-left} = 1.3$, $M_{up} = 3.1$, $M_{left} = 1.7$, and p = 1.5, by the definition $S(A_i, B_j) = 0.1$:

1.3	3.1
1.7	max(1.3 + 0.1,3.1 − 1.5,1.7 − 1.5) = 1.6

The final value of the $M_{n,m}$ (bottom-right of the matrix M) element will be the best score of the alignment algorithm. Finally, in order to find the actual alignment, it is necessary to backtrack on the path in the inverse direction from $M_{n,m}$ to $M_{0,0}$ [20].

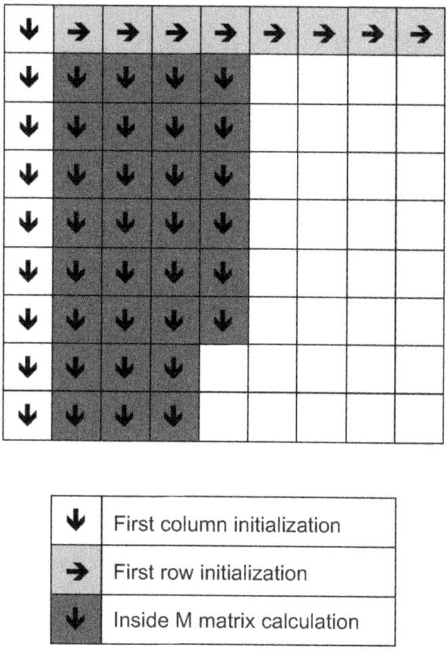

↓	First column initialization
→	First row initialization
↓	Inside M matrix calculation

Figure 20. Needleman-Wunsch M-matrix calculation.

This algorithm will always find the best solution, but ambiguity is in the similarity matrix definition and in the gap definition, which may be well defined or not. The time of computation is proportional to $n * m * t_s$, where t_s is the time needed for the pair similarity calculation ($S(A_i, B_j)$) [20].

3.2.4 *Other alignment methods*

Several text aligners implementing various alignment methods are currently available, including Bleualign,[12] which is an open source project developed

---

[12] https://github.com/rsennrich/bleualign

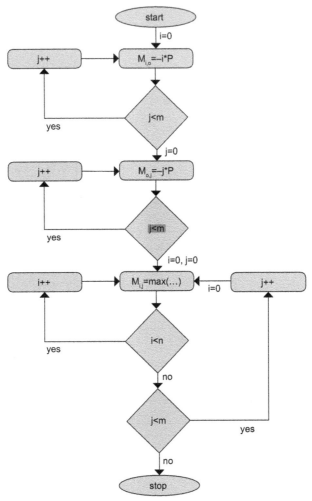

Figure 21. Diagram of the M matrix calculation without GPU optimization.

by the University of Zurich. In addition to parallel texts, Bleualign requires a translation of one of the texts. It uses the length-based Bilingual Evaluation Understudy (BLEU) similarity metric to align the texts [52].

The Hunalign[13] tool is another open source tool, developed by the Media Research Center. Based on Gale and Church's method, it uses sentence lengths and, optionally, a dictionary to align texts in two languages strictly

---

[13] http://mokk.bme.hu/resources/hunalign/

on a sentence level. In the presence of a dictionary, Hunalign combines the dictionary information with Gale-Church sentence-length information (It works on the principle that equivalent sentences should roughly correspond in length.) [201]. In the absence of a dictionary, it first falls back to sentence-length information and then builds an automatic dictionary based on this alignment. Then it realigns the text in a second pass, using the automatic dictionary [23].

Like most sentence aligners, Hunalign does not deal well with changes in sentence order [51]. It is unable to come up with crossing alignments, i.e., segments A and B in one language corresponding to segments B' A' in the other language.

ABBYY Aligner[14] is a commercial product developed by the ABBYY Group. This product reportedly uses proprietary word databases to align text portions of sentences based on their meaning [49].

Unitex Aligner[15] is an open source project primarily developed by the University of Paris-Est Marne-la-Vallée (France) [50]. It uses the XAlign tool [53], which uses character lengths at the paragraph and sentence level for text alignment [54].

Dyer's Fast Align algorithm applies a log-linear reparameterization to IBM Model 2's alignment model in order to solve the problem of Model 2 overfitting data and avoid the assumption of IBM Model 1 that all alignments are equally probable. The algorithm is designed to be fast to train and to compute text alignments [74].

---

[14] http://www.abbyy.com/aligner/
[15] http://www-igm.univ-mlv.fr/~unitex/

4

Author's Solutions to PL-EN Corpora Processing Problems

In this chapter, the author's solutions to some of the problems posed by this research and their implementation are described. First, improvements to the alignment method used in Yalign [19] are presented. In addition, a new method for mining parallel data from comparable corpora using a pipeline of tools is described. The adaptation of the BLEU evaluation metric for the needs of PL-EN translation is presented. Lastly, the author's method for aligning and filtering corpora at the sentence level is discussed.

4.1 Parallel data mining improvements

In this research, methodologies that obtain parallel corpora from data sources that are not sentence-aligned, such as noisy and parallel or comparable corpora, are presented. For this purpose, a set of specialized tools for obtaining, aligning, extracting and filtering text data was used. The tools were combined in a pipeline that allowed completion of the task in an integrated manner. The results of initial experiments on text samples obtained from Wikipedia and Euronews web pages are presented. Wikipedia was chosen as a data source because of the large number of documents that it provides (1,047,423 articles in PL Wiki and 4,524,017 in EN, at the time of writing). Furthermore, Wikipedia contains not only comparable documents, but also some documents that are translations of each other. The quality of the SMT approach used was measured by improvements in translations using BLEU measure.

For the experiments in statistical speech translation, the TED talks domain that was prepared for the IWSLT 2014 evaluation campaign by the FBK[16] was chosen. This domain is very wide and covers many unrelated subject areas. The data contains almost 2.5 M untokenized words [15]. Narrower domains were selected from the data for use in experiments. The first parallel corpus is composed of PDF documents from the European Medicines Agency (EMEA) and medicine leaflets [16]. The Polish data was a corpora created by the EMEA. Its size was about 80 MB, and it included 1,044,764 sentences built from 11.67 M untokenized words. The vocabularies consisted of 148,170 unique Polish words forms and 109,326 unique English word forms. The disproportionate vocabulary sizes are also a challenge, especially in translation from English to Polish.

The second corpus was extracted from the proceedings of the European Parliament (EUP) by Philipp Koehn (University of Edinburgh) [14]. In addition, experiments on the Basic Travel Expression Corpus (BTEC), a multilingual speech corpus containing tourism-related sentences similar to those usually found in phrase books for tourists going abroad, were also conducted [17]. Lastly, a large corpus obtained from the OpenSubtitles.org web site was used as an example of human dialogs. Table 4 provides details on the number of unique tokens (TOKENS), as well as the number of bilingual sentence pairs (PAIRS).

Table 4. Corpora specification.

CORPORA	PL TOKENS	EN TOKENS	PAIRS
BTEC	50,782	24,662	220,730
TED	218,426	104,117	151,288
EMEA	148,230	109,361	1,046,764
EUP	311,654	136,597	632,565
OPEN	1,236,088	749,300	33,570,553

The solution can be divided into three main steps. First, data is collected, then it is aligned at the article level, and finally the aligned results are mined for parallel sentences. The last two steps are not trivial, because there are great disparities between Wikipedia documents. Based on Wikipedia statistics, it is known that articles on the PL Wiki contain an average of 379 words, whereas articles on the EN Wiki contain an average of 590 words. This is most likely why sentences in the raw Wiki corpus are mostly misaligned, with translation

---

[16] https://wit3.fbk.eu/

lines whose placement does not correspond to any text lines in the source language. Moreover, some sentences have no corresponding translations in the corpus at all. The corpus might also contain poor or indirect translations, making alignment difficult. Thus, alignment is crucial for accuracy. Sentence alignment must also be computationally feasible to be of practical use in various applications.

The procedure that was applied in this monograph starts with a specialized web crawler. Because the PL Wiki contains less data, with almost all its articles having corresponding EN Wiki articles, the program crawls the data starting from the non-English site. The crawler can extract and save bilingual articles of any language supported by Wikipedia. The tool requires archives of at least two Wikipedia data sets on different languages and information about language links between the articles in the data sets.

A web crawler was implemented and used for Euronews.com. This web crawler was designed to use the Euronews.com archive page. In the first phase, the crawler generates a database of parallel articles in two selected languages, in order to obtain comparable data. The Wikipedia articles were analyzed offline using current Wikipedia dumps.[17]

Before a mining tool processes the data, it must be prepared. First, all the data is saved in a local database implemented using MongoDB.[18] Second, the tool aligns article pairs and removes articles that do not exist in both languages from the database. These topic-aligned articles are filtered to remove any HTML tags, XML tags or noisy data (tables, references, figures, etc.). Finally, bilingual documents are tagged with a unique ID as a topic-aligned, comparable corpus.

To extract the parallel sentence pairs, two different strategies were employed. The first strategy uses methodology inspired by the Yalign Tool,[19] and the second is based on a pipeline of specialized tools adapted for Polish language needs. The MT results presented in this chapter were obtained using the first strategy. This decision was motivated by the quality of Yalign documentation, experience with the algorithms, optimizations that were applied to it within the scope of this monograph, the computational feasibility of an improved Yalign method, and the unknown outcome of experiments with the tools pipeline described in Section 4.5.1. The second method is still under development; nevertheless, the initial results are promising

---

[17] https://dumps.wikimedia.org/
[18] https://www.mongodb.org/
[19] https://github.com/machinalis/yalign

and worth mentioning, especially since there are many opportunities for improvement.

4.2 Multi-threaded, tuned and GPU-accelerated Yalign

To improve the performance of Yalign and make it computationally feasible for large-scale data mining, a solution that supplies the Yalign tool with articles from the database within one session, with no need to reload the classifier each time, was developed. The developed solution also facilitated multi-threading and decreased the mining time by a factor of 6.1 (using a 4-core, 8-thread i7 CPU). The alignment algorithm was also reimplemented for better accuracy and to leverage the power of GPUs for additional computing requirements. The author's tuning algorithm was also implemented.

4.2.1 Needleman-Wunsch algorithm with GPU optimization

In the Needleman-Wunsch algorithm that has been optimized for GPUs, the structure of the algorithm remains the same, but it differs in the calculation of the M matrix elements. This calculation is the step to which multi-threading optimization is applied. Those operations are small enough to be processed by an enormous number of GPUs. The idea is to compute all elements in an diagonal in parallel, always starting from the top-left and proceeding to the bottom-right [20]. According to [194], the speed up in plain sequence alignment should be up to 8x. In this monograph, such a large speedup was not obtained, probably because of a bottleneck due to a need for data transfer to the GPU and the large size of the comparison matrix.

An example is presented in Table 5 and Figure 22. Imagine that the fourth diagonal elements of the M matrix ($M_{ad(4)}$) must be computed and that all the ($M_{ad(4)}$) elements are calculated in parallel threads. To make this calculation, the following equation is used:

$$max(\ldots) = max(M_{up-left} + S(A_{current}, B_{current}), M_{up} - p, M_{left} - p) \qquad (62)$$

Table 5. Matrix estimation example.

$M_{0,0}$	$M_{0,1}$	$M_{0,2}$	$M_{ad(4),3} = max(\ldots)$
$M_{1,0}$	$M_{1,1}$	$M_{ad(4),2} = max(\ldots)$	...
$M_{2,0}$	$M_{ad(4),1} = max(\ldots)$	...	...
$M_{ad(4),0} = max(\ldots)$	...	...	...

Next, the parallelization of the fifth diagonal will start, and so forth.

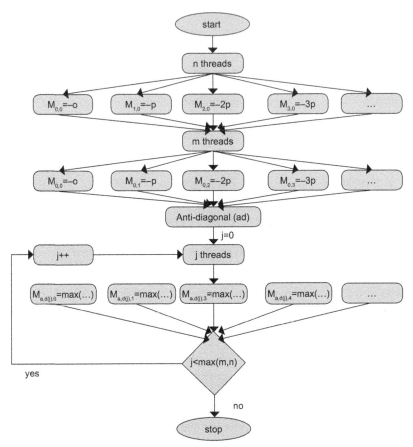

Figure 22. Diagram of the M matrix calculation with GPU optimization.

4.2.2 *Comparison of alignment methods*

The two NW algorithms, the existing algorithm and the algorithm improved through GPU optimization, are conceptually identical, but the second has an advantage in efficiency, depending on the hardware, up to *max(n, m)* times. However, the results of the A* algorithm, if the similarity calculation and the gap penalty are defined as in the NW algorithm, will be the same only if there is an additional constraint on paths: Paths cannot go upward or leftward in the M matrix. Yalign does not impose these additional conditions, so in some scenarios, repetitions of the same phrase may appear. In fact, every time the algorithm decides to move up or left, it is coming back into the second and first sequence, respectively.

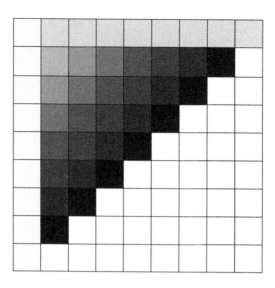

	First column initialization in parallel threads
	First raw initialization in parallel threads
	1st anti-diagonal calculation in parallel threads
	2nd anti-diagonal calculation in parallel threads
	3rd anti-diagonal calculation in parallel threads
	...

Figure 23. Needleman-Wunsch M-matrix calculation with parallel threads.

An example of an M matrix without constraints is presented in Table 6:

Table 6. M matrix calculation without constraints.

	a	d	e	g	f
a	X				
d		X			
c	X				
d		X			
e			X	X	X

The alignment in this scenario is:

a, d, a, d, e, g, f

a, d, c, d, e, −, −

In the same problem, the NW would react as presented in Table 7:

Table 7. M matrix calculation with NW.

	a	d	e	g	f
a	X				
d		X			
c		X			
d		X			
e			X	X	X

The alignment using NW would be:

a, d, −, −, e, g, f

a, d, c, d, e, −, −

Because of the lack of constraints, repetitions were created that visualized the imperfection of the A* algorithm implemented in the Yalign tool. The Yalign approach can save much time in a large data set of phrases, because the time of a single pair similarity calculation is negligible, but to obtain the same results as the NW method, some constraints must be inserted. This is not very easy, since Yalign uses an external package for the A* alignment that does not allow the required search path constraints. Many sentences may be misaligned or missed during the alignment, especially when the texts being analyzed are of different lengths and have vocabularies rich in synonyms. Some sentences can simply be skipped while checking for alignment. Such an issue occurs in Wikipedia articles for PL-EN pairs. That is why NW with GPU optimization is the most suitable algorithm. In this research, a comparison was made using all three approaches.

4.3 Tuning of Yalign method

Recall from Chapter 3 that the quality of alignments is defined by a tradeoff between precision and recall. The solution implemented in this research also introduces a tuning algorithm for these parameters, which allows better adjustment of them.

The tuning script consists of three main parts: An article parser, an aligner and a tuning algorithm. It requires some software and Python packages to be installed before using it: The original Yalign, MongoDB, python-pymongo, ipython, python-nose and python-werkzeug.

The Wikipedia article parser is a pre-processing tool, developed by the author, that extracts articles from Wikipedia as plain text data and prepares it for use with Yalign. It matches bilingual article pairs that are later processed by Yalign. The parser works in two steps. First, it extracts articles from Wikipedia data sets. Second, it saves the extracted articles to a local database (Mongo DB). Alignment at the article level is also done during this phase.

Before using the parser, Wikipedia data sets and language links should be downloaded and extracted to the same directory (directory should contain *.xml and *.sql files). For each language, two dump files should be downloaded, for example:

PL:

* http://dumps.wikimedia.org/plwiki/latest/plwiki-latest-pages-articles. xml.bz2

* http://dumps.wikimedia.org/plwiki/latest/plwiki-latest-langlinks.sql.gz

EN:

* http://dumps.wikimedia.org/enwiki/latest/enwiki-latest-pages-articles. xml.bz2

* http://dumps.wikimedia.org/enwiki/latest/enwiki-latest-langlinks.sql.gz

The article parser has a "main language" option. This option is used to define a main language, so that articles in a language are also extracted in other languages only if they exist in the main language. For example, if the main language is PL, then the article extractor will first extract all articles for PL, then extract articles in other languages, but only if such articles exist in the PL Wikipedia. This feature reduces space and computing requirements.

The aligner takes article pairs for a given language pair, aligns their text, and saves the parallel corpora to N files. The option "–s" can be used to limit the number of symbols in the file (by default, the size is 50,000,000 symbols, which has a size of 50–60 MB). This process, even after performance improvements that were made in this research, takes a very long time. For a PL-EN pair on the Intel Core i7, the process took about one month. That is why, by default, the aligner tries to continue aligning where it was stopped, in case of a computer crash.

To perform tuning, it is necessary to extract random article samples from the corpus. The script "save_plain_text.py" can be used to save articles in plain text format. It accepts a path for saving articles, the languages of articles to be saved and the source of articles (Euronews, Wikipedia, etc.). Such articles must be manually aligned by humans. It is essential to leave the order in which they appeared in the original text unmodified and to remove poor or incorrect alignments. Based on such information, the tuning script tries, naively by random parameter selection, to find values for which Yalign output is as similar to that of a human as possible. When the tuning script completes, it will produce a csv file, with the first column containing threshold, the second column containing penalty, and the third column containing similarity for these parameters. Similarity is a percentage value of how the automatically-aligned file resembles the human-aligned one. A Needleman-Wunsch algorithm is used for alignment and comparison of human made work with automatic script results.

Analysis was performed for each of five domains of interest to check how the tuning algorithms cope with proper adjustment of the parameters. Table 8 shows the results of this experiment.

Table 8. Improvements in quantity of mining.

Classifier	Improvement in %
BTEC	13.2
TED	12.5
EMEA	17
EUP	5
OPEN	11.4

For the testing purposes, 100 random article pairs were taken from the Wikipedia comparable corpus and aligned by a human translator. Second, a tuning script was run using classifiers trained on the previously-described text domain. A percentage change in quantity of mined data was calculated for each classifier.

4.4 Minor improvements in mining for Wikipedia exploration

The improvements discussed in the previous chapters mostly deal with heuristics used in the Yalign tool and can be applied to any bilingual textual data. Such an approach would also improve mining within Wikipedia. However, Wikipedia has many additional sources of cross-lingual dependencies that can be used. First, the topic domain of Wikipedia cannot be constrained to one specific domain; it covers almost any topic. This is the reason why its articles often contain complicated or rare vocabulary, and why statistical mining methods may skip some parallel sentences. The solution to this problem

Artykuł Dyskusja
Bocian biały [edytuj]
Bocian biały (Ciconia ciconia) –
WIKIPEDIA
The Free Encyclopedia
Article Talk
White stork
From Wikipedia, the free encyclopedia

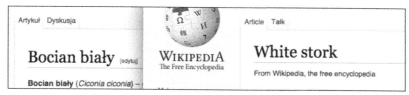

Figure 24. Example of bi-lingual Wikipedia page title.

might be extraction of a dictionary using the article titles from Wikipedia. According to [4], the precision of such a dictionary can as high as 92%. This dictionary can be used not only for the extension of parallel corpora but also in the classifier training phase.

Second, the textual descriptions of pictures contain good quality parallel phrases. It is possible to obtain them through picture analysis and hyperlinks to the pictures. The same goes for any figures, tables, maps, audio, video or any other multimedia content from Wikipedia. Unfortunately, not all information can be extracted from Wikipedia dumps; a web crawler is required for this task.

Młody bocian żerujący na owadach
A juvenile feeding on an insect

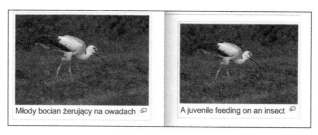

Figure 25. Example of bi-lingual picture description.

The good quality Wikipedia articles are well referenced. It is more likely for sentences to be cross-lingual equivalents if they are referenced with the same publication. Such analysis joined with other comparison techniques can lead to better accuracy in parallel text recognition.

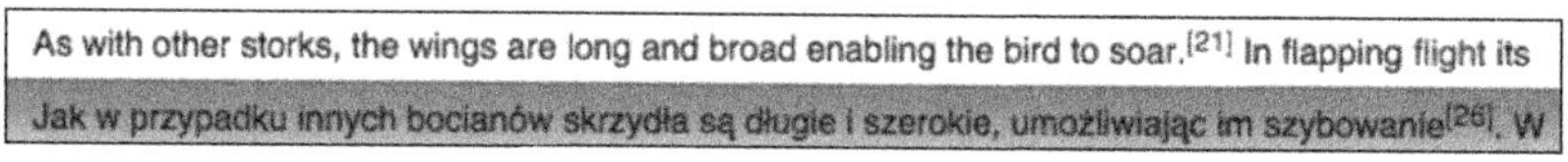

As with other storks, the wings are long and broad enabling the bird to soar.[21] In flapping flight its
Jak w przypadku innych bocianów skrzydła są długie i szerokie, umożliwiając im szybowanie[26]. W

Figure 26. Example of bi-lingual reference sentences.

Unfortunately, Wikipedia articles are developed separately for each language by many authors. In the following example, the parallel sentence in one language refers to reference number 21 and in another, to reference 26.

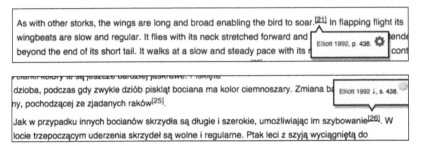

As with other storks, the wings are long and broad enabling the bird to soar.[21] In flapping flight its wingbeats are slow and regular. It flies with its neck stretched forward and [Elliott 1992, p. 438. ⚙] end< beyond the end of its short tail. It walks at a slow and steady pace with its n cont

dzioba, podczas gdy zwykle dziób piskląt bociana ma kolor ciemnoszary. Zmiana b[Elliott 1992 l, s. 438.] ny, pochodzącej ze zjadanych raków[25].

Jak w przypadku innych bocianów skrzydła są długie i szerokie, umożliwiając im szybowanie[26]. W locie trzepoczącym uderzenia skrzydeł są wolne i regularne. Ptak leci z szyją wyciągniętą do

Figure 27. Example of cross-lingual reference.

This is why it is necessary not only to compare numbers but to analyze the references themselves.

In addition to references, it is important to analyze names, dates, numbers, etc., because it can indicate the presence of parallel data.

In some cases, it would be useful to have an option to align more than two languages. The author's tool, unlike Yalign, can align any number of languages and build parallel, cross-lingual corpora from them.

Because of the need to use a web crawler, this tool version was evaluated only using 1,000 randomly-selected articles. It would take too long to build an entire corpora using it without access to many proxy servers and Internet connections. It is not only required to crawl about 500,000 articles for PL-EN pair, but also many links present on each page. First, the data was crawled, next aligned using the standard version of Yalign (Y in Table 9), and then aligned using the modified version described here (YMOD in Table 9). Lastly, single words were extracted and counted (DICT in Table 9) and also manually analyzed to verify how many of them could be considered correct translations (DICTC in Table 9).

Table 9. Results of mining after improvements.

Y	4,192
YMOD	5,289
DICT	868
DICTC	685

The results mean that the improved method using additional information sources mined an additional 1,097 parallel alignments. Among them, it is possible to identify 868 single words, which means that, in fact, 229 new sentences were obtained. The growth in obtained data was 5.5%. After manual analysis of the dictionary, 685 words were identified as proper translations. This means that the accuracy of the dictionary was about 79%. In summary, the author's method, though time-consuming, may produce additional results that would greatly help when dealing with text domains with insufficient textual resources.

4.5 Parallel data mining using other methods

This section describes a methodology for mining bi-data from previously obtained comparable corpora. The task is motivated by highly practical reasons. Non-parallel multilingual data exist in far greater quantities than parallel corpora, but parallel sentences are a much more useful resource. As previously described, a parser was used for building subject-aligned comparable corpora from Wikipedia articles. The method introduced for extracting parallel sentences that are filtered out from noisy or just comparable sentence pairs is presented in this part of the research. Implementation and adjustment of specialized tools for this task, as well as training and adaptation of a machine translation system that supplies a filtering tool with additional information about the similarity of comparable sentence pairs, are also described. In addition, initial results of the method based on analogies are also presented.

4.5.1 The pipeline of tools

The new procedure starts with a specialized web crawler, similar to the Yalign method. The procedure is presented in Figure 28. Because the PL Wiki contains less data, for which almost all articles have corresponding EN Wiki articles, the program crawls data starting from the non-English site first. It is a language-independent solution. The crawler can obtain and save bilingual articles of any language supported by Wikipedia.

Figure 28. Tools pipeline.

First, the data is saved in HTML files, and then it is topic-aligned. To narrow the search field to specific in-domain documents, it is necessary to give the crawler the first link to the article in a domain. Then the program will automatically obtain other topic-related documents, based on Interwiki links and references. Narrowing the search domain not only helps adjust the output to specific needs, but also narrows the vocabulary, which makes the alignment task easier. After obtaining HTML documents, the crawler extracts plain text from them and cleans the data. Tables, URLs, figures, pictures, menus, references and other unnecessary data are removed. Finally, bilingual documents are tagged with a unique ID for the topic-aligned comparable corpus.

A two-level sentence alignment method that prepares a dictionary for itself is proposed in this research. The Hunalign[20] tool is first used to match bilingual sentences. Its input is tokenized and sentence-segmented. The Hunalign option that automatically builds a dictionary was used [23].

To cope with Hunalign's inability to determine crossing alignments and filter out poor bilingual sentence pairs, a special tool was implemented [24]. The filtering tool is dependent on the translation engine. Since the best results can be obtained with specialized translation systems, PL-EN parallel data in various domains were extracted from the OPUS project and used for training a specialized machine translation system.

To improve its performance, the system was adapted to Wikipedia using a dump of all English articles as a language model. The final training corpora comprised 36,751,049 sentences, and the language model was composed of 79,424,211 sentences. The unique token count was 3,209,295 in the Polish side of the corpora; 1,991,418 in the English side; and 37,702,319 in the language model. The same system settings were used as for the baseline systems.

4.5.2 Analogy-based method

A method based on sequential analogy detection was developed in cooperation with Emilia Rejmund of the Polish-Japanese Academy of Information Technology [218]. Based on a parallel corpus, it detects analogies that exists in both languages. To enhance the quality of identified analogies, clusters of sequential analogies are sought.

However, this research (based on the Wikipedia corpus) shows that it is both extremely difficult and computationally intensive to seek out clusters of higher orders. Therefore, this research was restrained to simple analogies, such as A is to B in the same way as C to D.

$$A:B::C:D$$

Such analogies are found using a distance calculation. The method seeks sentences such that:

$$dist(A,B)=dist(C,D)$$

and

$$dist(A,C)=dist(B,D)$$

An additional constraint was added that requires the same relation of occurrences of each character in the sentences. For example, if the number of occurrences of the character "a" in sentence A is equal to x and equal to y in sentence B, then the same relation must occur in sentences C and D.

---

[20] http://mokk.bme.hu/resources/hunalign/

The Levenshtein metric was used in this part of the present research for distance calculation. A trial application of this metric was made, applying it directly to the characters in the sentence, or considering each word in the sentence as an individual symbol, and calculating Levenshtein distance between symbol-coded sentences. The latter approach was employed, due to the fact that this method was tested earlier on the Chinese and Japanese languages [195], which use symbols to represent entire words.

After clustering, data from clusters are compared to each other in order to find similarities between them. For the four sentences:

A:B::C:D

The algorithm looks for such E and F that:

C:D::E:F and E:F::A:B

However, none were found in our corpus; therefore, the experiments were constrained to small clusters with two pairs of sentences. Matching sentences from the parallel corpus were identified in every cluster. This allowed the generation of new, similar sentences, which were present in the training corpus. For each of the sequential analogies that were identified, a rewriting model is constructed. This was achieved by string manipulation. Common prefixes and suffixes for each of the sentences were calculated using the Longest Common Subsequence (LCS) method [196].

An example of the rewriting model (prefix and suffix are shown in bold) is:

Poproszę koc i poduszkę. ⇔ *A blanket and a pillow,* ***please.***

Czy mogę poprosić o śmietankę i cukier? ⇔ ***Can I have*** *cream and sugar?*

The rewriting model consists of a prefix, a suffix, and their translation. It is possible to construct a parallel corpus from a non-parallel, monolingual source. Each sentence in the corpus is tested for a match with the model. If the sentence contains a prefix and a suffix, it is considered a matching sentence.

Poproszę bilet. ⇔ *A unknown,* ***please.***

In the matched sentence, some of the words remained untranslated, but the general meaning of the sentence is conveyed. Remaining words may be translated word-for-word, while the translated sentence remains grammatically correct.

bilet ⇔ ticket

Substituting unknown words with translated ones allowed the creation of a parallel corpus entry.

Poproszę bilet. ⇔ *A ticket,* ***please.***

As a result of the sequential analogy detection-based method, 8,128 models were mined from the Wikipedia comparable corpus. This enabled generation of 114,000 new pairs of sentences that could be used as an extension to a parallel corpus. Sentences were generated from a Wikipedia comparable corpus that is basically an extract of Wikipedia articles. It contained articles in Polish and English on the same topic, but sentences were not aligned in any particular way. The rewriting models were used to match sentences from each Polish article to sentences in the English article. Whenever a model could be successfully applied to a pair of sentences, this pair was considered to be parallel, resulting in the generation of a quasi-parallel corpus ("quasi" since sentences were aligned artificially using the approach described above). Those parallel sentences can be used to extend parallel corpora in order to improve the quality of machine translation.

4.6 SMT metric enhancements

This section describes enhancements made to the BLEU metric and the evaluation of the enhanced SMT metric.

4.6.1 Enhancements to the BLEU metric

When human translators translate a text, they often use synonyms, different word orders or style and other similar variations. Here, an SMT evaluation technique that enhances the BLEU metric to consider such variations is proposed.

In particular, the enhanced metric rewards synonyms and rare word translations, while modifying the calculation of cumulative scores. The enhanced metric rewards matches of synonyms, as does BLEU, since the correct meaning is still conveyed.

Consider the test phrase "this is a exam" and the reference phrase "this is a quiz." The BLEU score is calculated as follows:

$$BLEU = \frac{(1+1+1+0)}{4} = \frac{3}{4} = 0.75 \tag{63}$$

BLEU does not count the word "exam" as a match, because it does not find it in the reference phrase. However, this word is not a bad choice. In the proposed method, the algorithm scores the synonym "exam" higher than zero and lower than the exact word "quiz."

To do this, the evaluation algorithm tries to find the synonyms of each word in a test phrase. It checks for an exact word match and for all test phrase synonyms to find the closest words to the reference. For example, for the phrases:

Test: "this is a exam"

Reference: "this is a quiz"

"exam" has some synonyms, e.g., "test", "quiz" and "examination."

The algorithm checks each synonym (taken from WordNet[21]) in the reference. If a synonym has a greater number of matches in the reference, it is replaced with the original word.

The default BLEU algorithm was applied to the modified test phrase and reference phrase, with one difference. The default BLEU algorithm scores this new test phrase as 1.0, but it is known that the original test phrase was "this is a exam." So, the scoring algorithm should give a score higher than 0.75 but less than 1.0 to the test phrase.

During the BLEU evaluation, the program checks each word for an exact match. If the word is a synonym and not an exact match, it does not give a full score to that word. The score for a synonym will be the default BLEU score for an original word multiplied by a constant (synonym-score).

For example, if this constant equals 0.90, the new score with synonyms is:

$$(1 + 1 + 1 + 0.9)/4 = 3.9/4 = 0.975$$

With this algorithm, synonym scores for all n-grams are obtained, because in the 2-grams, there are "a quiz" and in the 3-grams, "is a quiz" in both the test and reference phrases.

Additionally, the algorithm gives extra points to rare word matches. First, it obtains the rare words found in the reference corpus. If all distinct words of the reference are sorted in their repetition order (descending), the last words in this list are rare words. The algorithm takes a specific percentage of the whole sorted list as rare words (rare-words-percent).

When the default BLEU algorithm tries to score a word, if this word is in the rare word list, the score is multiplied by a constant (rare-words-score). This action applies to all n-grams. So, if rare words are found in a 2-gram, the algorithm increases the score for this 2-gram. For example, if the word "roman" is rare, then the 2-gram "roman empire" gets an increased score.

The algorithm is careful that the score of each sentence falls within the range of 0.0 and 1.0. The cumulative score of the algorithm combines default BLEU scores using logarithms and exponentials as follows:

1. Initialize s = 0

2. For each ith-gram:

---

[21] https://wordnet.princeton.edu/

a. $s = s + log(B_i)$

b. $C_i = exp(s/i)$

where B_i is the default BLEU score, and C_i is the cumulative score.
   In addition, knowing that:

$$exp(log(a) + log(b)) = a * b \qquad (64)$$

and:

$$exp(log(a)/b) = a \wedge (1/b) \qquad (65)$$

simplifies the calculation. For example, for i = 1 to 4:

$$C1 = B1$$
$$C2 = (B1 * B2) \wedge (1/2) \qquad (66)$$
$$C3 = (B1 * B2 * B3) \wedge (1/3)$$
$$C4 = (B1 * B2 * B3 * B3) \wedge (1/4)$$

Assuming that:

$$B1 = 0.70$$
$$B2 = 0.55$$
$$\qquad (67)$$
$$B3 = 0.37$$
$$B4 = 0.28$$

then:

$$C1 = 0.70$$
$$C2 = 0.62$$
$$\qquad (68)$$
$$C3 = 0.52$$
$$C4 = 0.44$$

The length score (brevity penalty) in the algorithm is calculated as:

$$len_{score} = min(\frac{0.0,\, 1 - ref_{length}}{test_{ngrams}}) \qquad (69)$$

and cumulatively:

$$exp(\frac{score}{i} + len_{score}) \qquad (70)$$

4.6.2 *Evaluation using enhanced BLEU metric*

Experiments were conducted in order to compare the performance of the enhanced SMT evaluation metric with that of the most popular metrics for SMT between Polish and English: BLEU, NIST, TER and METEOR. The data set used for the experiments was the EMEA) parallel corpus [35].

Table 10 shows the results of Polish-to-English translation experiments. Table 11 shows the results of English-to-Polish translation experiments. The EBLEU column is an evaluation of the new metric.

Table 10. Polish-to-English translations results.

EXP NO	EBLEU	BLEU	NIST	TER	MET	RIBES
00	70.42	70.15	10.53	29.38	82.19	83.12
01	63.75	64.58	9.77	35.62	76.04	72.23
02	70.85	71.04	10.61	28.33	82.54	82.88
03	70.88	71.22	10.58	28.51	82.39	83.47
04	76.22	76.24	10.99	24.77	85.17	85.12
05	70.94	71.43	10.60	28.73	82.89	83.19
06	73.10	71.91	10.76	26.60	83.63	84.64
07	70.47	71.12	10.37	29.95	84.55	76.29
08	71.78	71.32	10.70	27.68	83.31	83.72
09	70.65	71.35	10.40	29.74	81.52	77.12
10	71.42	70.34	10.64	28.22	82.65	83.39
11	73.11	72.51	10.70	28.19	82.81	80.08

Table 11. English-to-Polish translations results.

EXP NO	EBLEU	BLEU	NIST	TER	MET	RIBES
00	66.81	69.18	10.14	30.90	79.21	82.92
01	58.28	61.15	9.19	39.45	71.91	71.39
02	67.24	69.41	10.14	30.90	78.98	82.44
03	66.33	68.45	10.06	31.62	78.63	82.70
04	72.00	73.32	10.48	27.05	81.72	84.59
05	67.31	69.27	10.16	30.80	79.30	82.99
06	66.64	68.43	10.07	31.27	78.95	83.26
07	66.41	67.61	9.87	33.05	77.82	77.77
08	66.64	68.98	10.11	31.13	78.90	82.38
09	67.30	68.67	10.02	31.92	78.55	79.10
10	66.76	69.01	10.14	30.84	79.13	82.93
11	66.66	67.47	9.89	33.32	77.65	75.19

The Wilcoxon signed-rank test was used to determine whether or not the differences between metrics were statistically significant. Table 12 shows the significance results for Polish-to-English translation, and Table 13 for English-to-Polish translation.

Table 12. Significance of EBLEU results versus other metrics (PL-EN).

METRIC	BLEU	NIST	TER	METEOR	RIBES
P-value:	0.843	0.000	0.000	0.000	0.000

The results presented in Tables 12 and 13 include one-tailed and two-tailed tests. There is no statistical difference between EBLEU and BLEU. The differences between EBLEU and the other columns are highly significant.

Table 13. Significance of EBLEU results versus other metrics (EN-PL).

METRIC	BLEU	NIST	TER	METEOR	RIBES
P-value:	0.000	0.000	0.000	0.000	0.000

The English-to-Polish translation results are significantly different from the EBLEU, and the results are highly significant. Even though the differences between BLEU and EBLEU are significant only in translation from English to Polish, such a result was anticipated, because the EBLEU metric was intended for evaluation of texts written in Polish.

Additionally, correlation was used to better assess the association among the metrics. Correlation measures the association among two or more quantitative or qualitative independent variables [17]. So, it was used here to estimate the association between metrics.

The correlation between two arrays of variables, X and Y, can be calculated using the following formula:

$$Corr(X,Y) = \frac{\sum (x - \bar{x})(y - \bar{y})}{\sqrt{\sum (x - \bar{x})^2 \sum (y - \bar{y})^2}} \tag{71}$$

The correlation output table for the metrics is:

Table 14. Correlation for Polish to English.

	EBLEU	BLEU	NIST	TER	METEOR	RIBES
EBLEU	1					
BLEU	0.9732	1				
NIST	0.9675	0.9158	1			
TER	−0.9746	−0.9327	−0.9909	1		
METEOR	0.8981	0.8943	0.8746	−0.8963	1	
RIBES	0.7570	0.6738	0.8887	−0.8664	0.6849	1

Table 14 shows that the NIST metric is more strongly correlated with EBLEU than with BLEU. The new metric shows a more negative correlation with TER than does BLEU. The new metric shows a stronger correlation with METEOR than does BLEU.

Figure 29 shows the data trends, as well as the association of different variables.

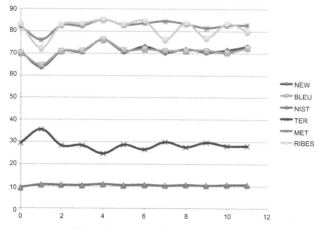

Figure 29. Association of metric values.

To confirm the results, it was of interest to determine if the same correlation would occur in translations from English to Polish. The results were determined, and an aggregation table was developed in which the results of both tables were merged. The aggregation is shown in Table 15.

Table 15. Aggregation for English-Polish.

	EBLEU	BLEU	NIST	TER	METEOR	RIBES
EBLEU	1					
BLEU	0.9657	1				
NIST	0.9762	0.9361	1			
TER	−0.9666	−0.9725	−0.9723	1		
METEOR	0.9615	0.9276	0.9653	−0.9411	1	
RIBES	0.8105	0.6989	0.9809	−0.9097	0.6849	1

Table 15 shows a stronger correlation between NIST and RIBES and the new metric than between NIST or RIBES and BLEU. The enhanced metric has a more negative correlation with TER than does BLEU. Lastly, the new metric has a stronger correlation with METEOR than does BLEU.

Finally, it was essential to confirm how statistically relevant the obtained results were. To check the correlation coefficiency, the asymmetric lambda measure of association $\lambda(C|R)$ was calculated. Asymmetric lambda is interpreted as the probable improvement in prediction of the column variable Y given knowledge of the row variable X (values given in table). Asymmetric lambda has the range:

$$0 \leq \lambda(C|R) \leq 1 \qquad (72)$$

It is computed as:

$$\lambda(C|R) = \frac{\sum_i r_i - r}{n - r} \qquad (73)$$

with:

$$var = \frac{n - \sum_i r_i}{(n-r)^3} \left(\sum_i r_i + r - 2\sum_i (r_i \mid l_i = l) \right) \qquad (74)$$

where:

$$Cr_i = \overset{max}{j}(n_{ij})$$
$$r = \overset{max}{j}(n_{.j}) \qquad (75)$$

For this purpose, IBM's SPSS (predictive analytics software)[22] tool was used [36]. In the experiments, the EBLEU results as a dependent variable (EBLEU is a function of the metrics variable) of every other metric were calculated. Using the interpretive guide for measures of association (0.0 = no relationship, ±0.0 to ±0.2 = very weak, ±0.2 to ±0.4 = weak, ±0.4 to ±0.6 = moderate, ±0.6 to ±0.8 = strong, ±0.8 to ±1.0 = very strong, and ±1.0 = perfect relationship), the lambda results were characterized as a very strong relationship if larger than 0.8 value [37]. Table 16 represents the strength of association between EBLEU and the other metrics.

Table 16. Association strength.

	BLEU*	NIST*	TER*	MET*	RIB**
Symmetric	0.973	0.918	0.957	0.975	0.978
EBLEU Dependent	0.988	0.870	0.957	1.000	1.000
* Dependent	0.958	0.957	0.958	0.958	0.957

The lambda results confirm that correlation is very strong for each metric. In the case of METEOR, there is perfect correlation.

---

[22] http://www-01.ibm.com/software/analytics/spss/

Lastly, the Spearman correlation [38] was evaluated. Its rank is often denoted by ρ (rho) or as r_s. It is a nonparametric measure of statistical dependence between two variables. It enables assessment of how well the relationship between two variables can be described using a monotonic function. If no repeated data values are found, then a perfect Spearman correlation of $+1$ or -1 occurs.

Pearson correlation is unduly influenced by outliers, unequal variances, non-normality and nonlinearity. This latter correlation is calculated by applying the Pearson[23] correlation formula to the ranks of the data rather than to the actual data values themselves. In so doing, many of the distortions that plague the Pearson correlation are reduced considerably.

Pearson correlation measures the strength of a linear relationship between X and Y. In the case of nonlinear but monotonic relationships, a useful measure is *Spearman's* rank correlation coefficient, *Rho*, which is a *Pearson's* type correlation coefficient computed on the ranks of the X and Y values. It is computed by the following formula:

$$Rho = \frac{[1 - 6\sum (d_i)^2]}{[n(n^2 - 1)]} \tag{76}$$

where:

d_i is the difference between the ranks of X_i and Y_i.

$r_s = +1$ if there is a perfect agreement between the two sets of ranks.

$r_s = -1$ if there is a complete disagreement between the two sets of ranks.

Spearman's coefficient, like any correlation calculation, is appropriate for both continuous and discrete variables, including ordinal variables. Table 17 shows the two-tailed Spearman's correlation for the EBLEU metric in the Correlation Coefficient row. The Sigma row represents the error rate (should be less than 0.05), and N is number of samples used in the experiment. Table 18 provides the results of Spearman's correlation for the BLEU metric.

The sigma value for the Spearman correlation indicates the direction of association between X (the independent variable) and Y (the dependent variable). If Y tends to increase when X increases, the Spearman correlation coefficient is positive. If Y tends to decrease when X increases, the Spearman correlation coefficient is negative. A Spearman correlation of zero indicates that there is no tendency for Y to either increase or decrease when X increases. The Spearman correlation increases in magnitude as X and Y come closer to being perfect monotonic functions of each other. When X and Y are perfectly monotonically related, the Spearman correlation coefficient is 1.

---

[23] http://onlinestatbook.com/2/describing_bivariate_data/pearson.html

Table 17. Spearman correlation for EBLEU.

	BLEU	NIST	TER	MET	RIB
Corr. Coefficient	0.950	0.943	−0.954	0.895	0.655
Sigma (2-tailed)	0.000	0.000	0.000	0.000	0.001
N	26	26	26	26	26

For example, −0.951 for TER and EBLEU shows a strong negative correlation between these values. Other results also confirm strong correlations between the measured metrics. Correlation between EBLEU and BLEU is 0.947, EBLEU and NIST is 0.940, EBLEU and TER is −0.951, and EBLEU and METEOR is 0.891. This shows strong associations between these variables. The results for the RIBES metric show moderate rather than very strong correlation.

Table 18. Spearman correlation for BLEU.

	EBLEU	NIST*	TER*	MET*	RIB*
Corr. Coefficient	0.950	0.915	−0.945	0.897	0.655
Sigma (2-tailed)	0.000	0.000	0.000	0.000	0.001
N	26	26	26	26	26

On the other hand, for the BLEU metric, the following correlation coefficient results were obtained: BLEU and NIST is 0.912, BLEU and TER is 0.939, and BLEU and METEOR is 0.897. This shows a strong association among the metrics as well, with EBLEU showing the strongest association. Low correlation for RIBES occurs for each kind of translation.

In this research, it was proved by measuring correlations that the author's enhanced BLEU metric is more trustworthy than normal BLEU. There are no deviations from the measurements from other metrics. Moreover, the new method of evaluation is more similar to human evaluation. In conclusion from the experiments, the new evaluation metric provides better precision, especially for Polish and other Slavic languages. As anticipated, the correlation between the new metric and RIBES is not very strong. The focus of the RIBES metric is word order, which is almost free in the Polish language. To be more precise, it uses rank correlation coefficients based on word order to compare SMT and reference translations. As word order is not strict in Polish, having a rather weak correlation with RIBES is a good indication.

The enhanced BLEU metric can deal with disparity of vocabularies between language pairs and the free-word order that occurs in some non-positional languages. The metric tool provides an opportunity for future adjustments in scoring. It allows changes in the proportions for which the BLEU score was enhanced. Because of this, the tool can be easily adapted to any language pair or specific experimental needs.

4.7 Alignment and filtering of corpora

A goal of the present research was to remove incorrect translations and noisy data from parallel corpora, in order to ultimately improve SMT performance. Before a parallel corpus can be used for any processing, the sentences must be aligned. Sentences in a raw corpus are often misaligned, resulting in translated lines whose placement does not correspond to the text lines in the source language. Moreover, some sentences may have no corresponding translation in the corpus at all. The corpus might also contain poor or indirect translations, making the alignment difficult and quality of the data poor. Thus, it is crucial to MT system accuracy [39]. Corpora alignment must also be computationally feasible to be of a practical use in various applications [40].

4.7.1 Corpora used for alignment experiments

The data set used for this research was the Translanguage English Database (TED) [15], provided by Fondazione Bruno Kessler (FBK) for IWSLT 2013. The vocabulary sizes of the English and Polish texts in TED are disproportionate. There are 41,684 unique English words and 92,135 unique Polish words. This also presents a challenge for SMT systems.

Additionally, a Wikipedia comparable corpus was built and used in this research. The corpus was prepared for the needs of the IWSLT 2014 Evaluation Campaign. In Table 19, a fragment of the Wikipedia corpora, which consists of 13 lines obtained with the original Yalign tool, is presented. There are many similar cases and fragments in the entire corpora that might be quite problematic during training of a machine translation system. First, the data is quite noisy, and the corpora contain redundant parallel lines consisting solely of numbers or symbols. In addition, it is easy to find improper translations, e.g., "U.S. Dept." is clearly not an accurate translation of the sentence "Na początku lat 30", which in Polish means "In the beginning of the 30s." In addition, some translations are too indirect or too distinct from each other. An example of such a pair is: "In all other cases it is true" and "W przeciwnym razie alternatywa zdań jest fałszywa", which in Polish means "Otherwise, the alternative of the sentences is false."

Despite these errors, the corpus contains good translations, and the problematic data can be removed.

Table 19. Example of poor-quality translations in the Wikipedia corpus.

Line	PL	EN
1	W przeciwnym razie alternatywa zdań jest fałszywa.	In all other cases it is true.
2	ISBN 8389795299.	ISBN 1-55164-250-6.
3	ISBN 8311068119.	ISBN 978-1-55164-311-3.
4	ISBN 8322924984.	ISBN 9780691155296.
5	ISBN 9788361182085.	ISBN 0-14-022697-4.
6	ASN.1 (skrót od „Abstract Syntax Notation One"—abstrakcyjna notacja składniowa numer jeden) jest to standard służący do opisu struktur przeznaczonych do reprezentacji, kodowania, transmisji i dekodowania danych.	Abstract Syntax Notation One (ASN.1) is a standard and notation that describes rules and structures for representing, encoding, transmitting, and decoding data in telecommunications and computer networking.
7	Standard ASN.1 określa jedynie składnię abstrakcyjną informacji, nie określa natomiast sposobu jej kodowania w pliku.	ASN.1 defines the abstract syntax of information but does not restrict the way the information is encoded.
8	Metody kodowania informacji podanych w składni ASN.1 zostały opisane w kolejnych standardach ITU-T/ISO.	X.683 | ISO/IEC 8824-4 (Parameterization of ASN.1 specifications)Standards describing the ASN.1 encoding rules:* ITU-T Rec.
9	pierwszego).	One criterion.
10	problemy nierozstrzygalne.	.".
11	Jeżeli dany algorytm da się wykonać na maszynie o dostępnej mocy obliczeniowej i pamięci oraz akceptowalnym czasie, to mówi się, że jest obliczalny.	I prefer to call it merely a logical-diagram machine ... but I suppose that it could do very completely all that can be rationally expected of any logical machine".
12	temperaturę) w optymalnym zakresie.	"Algorithmic theories"...
13	Na początku lat 30.	U.S. Dept.

For this part of the research, the Wikipedia corpus was extracted from a comparable corpus generated from Wikipedia articles. It was about 104 MB in size and contained 475,470 parallel sentences. Its first version was acknowledged as permissible data for the IWSLT 2014 evaluation campaign.

The TED Talks were chosen as a representative sample of noisy parallel corpora. The Polish data in the TED talks (about 15 MB) include almost 2 million words that are not tokenized. The transcripts themselves are provided as pure text encoded in UTF-8 format [27]. In addition, they are separated into sentences (one per line) and aligned in language pairs. However, some discrepancies in the text parallelism are present. These discrepancies are mainly repetitions of Polish text not included in the parallel English text.

Another problem was that the TED 2013 data contained many errors. This data set had spelling errors that artificially increased the dictionary size and made the statistics unreliable. A very large Polish dictionary [40], consisting of 2,532,904 different words, was extracted. Then, a dictionary consisting of 92,135 unique words forms was created from TED 2013 data. The intersection of those two dictionaries resulted in a new dictionary containing 58,393 words. This means that 33,742 words that do not exist in Polish (e.g., due to spelling errors or named entities) were found in TED 2013. This is 36.6% of the whole TED Polish vocabulary [18].

To verify this, a manual analysis of a sample of the first 300 lines from the TED corpus was conducted. It was found that there were 4,268 words containing a total of 35 kinds of spelling errors that occurred many times. But what was found to be more problematic was that there were sentences with odd nesting, such as:

Part A, Part A, Part B, Part B, for example:

"Ale będę starał się udowodnić, że mimo złożoności, Ale będę starał się udowodnić, że mimo złożoności, istnieją pewne rzeczy pomagające w zrozumieniu. Istnieją pewne rzeczy pomagające w zrozumieniu."

Some parts (words, full phrases, or even entire sentences) were duplicated. Furthermore, there are segments containing repetitions of whole sentences inside one segment. For instance:

Sentence A, Sentence A, for example:

"Zakumulują się u tych najbardziej pijanych i skąpych. Zakumulują się u tych najbardziej pijanych i skąpych."

or: Part A, Part B, Part B, Part C, for example:

"Matka może się ponownie rozmnażać, ale jak wysoką cenę płaci, przez akumulację toksyn w swoim organizmie - przez akumulację toksyn w swoim organizmie - śmierć pierwszego młodego."

The analysis identified 51 out of 300 segments that contained such mistakes. Overall, 10% of the sample test set contained spelling errors, and about 17% contained insertion errors. However, it must be noted that simply the first 300 lines were taken, while in the whole text there were places where even more problems occurred. So, to some extent, this confirms that there were problems related to the dictionary.

In addition, there were numerous untranslated English names, words and phrases present in the Polish text. There are also some sentences that originate from other languages (e.g., German and French). Some translations were simply incorrect or too indirect without sufficient precision, e.g., "And I remember sitting there at my desk thinking, Well, I know this. This is a great scientific discovery" was translated into "Pamiętam, jak pomyślałem: To wyjątkowe, naukowe odkrycie." The correct translation would be "Pamiętam jak siedząc przy biurku pomyślałem, dobrze, wiem to. To jest wielkie naukowe odkrycie."

Another serious problem discovered (especially for statistical machine translation) was that English sentences were translated in an improper manner.

There were four main problems:

1. Repetitions—part of the text is repeated several times after translation, for example:

   a. EN: Sentence A. Sentence B.

   b. PL: Translated Sentence A. Translated Sentence B. Translated Sentence B. Translated Sentence B.

2. Wrong usage of words—when one or more words used for the Polish translation slightly change the meaning of the original English sentence, for example:

   a. EN: We had these data a few years ago.

   b. PL (the proper meaning of the Polish sentence): We've been delivered these data a few years ago.

3. Indirect translations or usage of metaphors—when the Polish translation uses a different vocabulary to preserve the meaning of the original sentence, especially when the exact translation would result in a sentence that makes no sense. Many metaphors are translated this way.

4. Translations that are not precise enough—when the translated fragment does not contain all the details of the original sentence, but only its overall meaning is the same.

4.7.2 *Filtering and alignment algorithm*

A bi-sentence filtering and aligning tool was designed to find an English translation of each Polish line in a corpus and place it in the correct place in the English file. The tool assumes that each line in a text file represents one entire sentence. The aligner and filter are implemented in the same tool; they can cope with both tasks by proper adjustment of parameters.

The initial concept was to use the Google Translator Application Programming Interface (API) lines for which an English translation does not exist and also for comparison between the original and translated texts. The second concept was based on web crawling, using Google Translator, Bing Translator and the Babylon translator. These can work in a parallel manner to improve performance. In addition, each translator can work in many instances. This approach can also accommodate a user-provided translation file in lieu of crowd sourcing. Any machine translation system can be used as well, as a source of translations.

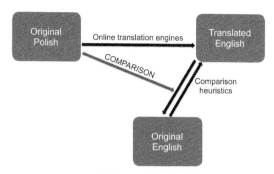

Figure 30. Comparison algorithm.

The strategy is to find a correct translation of each Polish line aided by Google Translator or another translation engine. The tool translates all lines of the Polish file (src.pl) with an SMT system and puts each translated line in an intermediate English translation file (src.trans). This intermediate translation helps in finding the correct line in the English translation file (src.en) and putting it in the correct position (Figure 30).

In reality, the actual solution is more complex. Suppose that one of the English data lines was chosen as the most similar line to a specific, translated line and that this line has a similarity rate high enough to be accepted as the translation. This line may be more similar to the next line of src.trans, such that the similarity rate of this selected line and the next line of src.trans is higher. For example, consider the sentences and their similarity rating in Table 20. The similarity was calculated using Python difflib library, which calculates similarity of the strings.[24]

Table 20. Example similarity ratings.

src.trans	src.en	Sim.
I go to school every day.	I like going to school every day.	0.60
I go to school every day.	I do not go to school every day.	0.70
I go to school every day.	We will go tomorrow.	0.30
I don't go to school every day.	I like going to school every day.	0.55
I don't go to school every day.	I do not go to school every day.	0.95
I don't go to school every day.	We will go tomorrow.	0.30

In this situation, "I do not go to school every day"-not "I go to school every day"-should be selected from src.en instead of "I don't go to school every day" from src.trans. So, considering the similarity of a selected line

---

[24] https://docs.python.org/2/library/difflib.html

with the next lines of src.trans enables making the best possible selection in the alignment process.

There are additional complexities that must be addressed. Comparing the src.trans lines with the src.en lines is not easy, and it becomes harder when usage of the similarity rate to choose the correct, real-world translation is required. There are many strategies to compare two sentences. It is possible to split each sentence into its words and find the number of common words in both sentences. However, this approach has some problems. For example, let us compare "It is origami" to these sentences: "The common theme what makes it origami is folding is how we create the form" and "This is origami."

With this strategy, the first sentence is more similar, because it contains all 3 words. However, it is clear that the second sentence is the correct choice. This problem can be solved by dividing the number of words in both sentences by the number of total words in the sentences. However, counting stop words in the intersection of sentences sometimes causes incorrect results. So, stop words are removed before comparing two sentences.

Another problem is that sometimes stemmed words can be found in sentences, for example, "boy" and "boys." Despite the fact that these two words should be counted as similar in two sentences, the words are not counted with this strategy.

The next comparison problem is the word order in sentences. In Python, there are other ways for comparing strings that are better than counting intersection lengths. The Python "difflib" library for string comparison contains a function that first finds matching blocks of two strings. For example, difflib can be used to find matching blocks in the strings "abxcd" and "abcd."

Difflib's "ratio" function divides the length of matching blocks by the length of two strings and returns a measure of the sequences' similarity as a float value in the range <0, 1>. This measure is 2.0 * M/T, where T is the total number of elements in both sequences, and M is the number of matches. Note that this measure is 1.0 if the sequences are identical, and 0.0 if they have nothing in common. Using this function to compare strings instead of counting similar words helps to solve the problem of the similarity of "boy" and "boys." It also solves the problem of considering the position of words in sentences.

Another problem in comparing lines is synonyms. For example, consider these sentences: "I will call you tomorrow"; "I would call you tomorrow." If it is essential to know if these sentences are the same, the algorithm should "know" that "will" and "would" can be used interchangeably.

The NLTK Python module and WordNet were used to find synonyms for each word, and these synonyms were used in comparing sentences. Using synonyms of each word, it was necessary to create multiple sentences from each original sentence, which is not computationally feasible. For example, suppose that the word "game" has the synonyms: "play", "sport", "fun", "gaming", "action" and "skittle." If used, for example, for the sentence "I do not like game", the following sentences are created: "I do not like play"; "I do not like sport"; "I do not like fun"; "I do not like gaming"; "I do not like action"; and "I do not like skittle" It is necessary to use each word in a sentence.

Next, the algorithm should try to find the best score by comparing all these sentences, instead of just comparing the main sentence. One issue is that this type of comparison takes a long time, because the algorithm needs to do many comparisons for each selection.

Difflib has other functions (in SequenceMatcher and Diff class) to compare strings that are faster than the described solution, but their accuracy is worse. To overcome all these problems and obtain the best results, two criteria should be considered: The speed of the comparison function and the comparison acceptance rate.

To obtain the best results, the script provides users with the ability to specify multiple functions with multiple acceptance rates. Fast functions with lower-quality results are tested first. If they can find results with a very high acceptance rate, the tool should accept their selection. If the acceptance rate is not sufficient, it can use slower but more accurate functions. The user can configure these rates manually and test the resulting quality to get the best results. All are well-described in documentation [44].

Some additional scoring algorithms were also implemented. These are more suited for comparing language translation quality. The BLEU and native implementations of TER and Character Edit Rate (CER) that use pycdec and RIBES were implemented, as well [120].

These algorithms were used to generate likelihood scores for two sentences, to choose the best one in the alignment process. For this purpose, the cdec and pycdec Python modules were used. The cdec module is a decoder, aligner and learning framework for statistical machine translation and similarly-structured prediction models. It provides translation and alignment modeling based on finite-state transducers and synchronous context-free grammars, as well as implementations of several parameter-learning algorithms [121, 122]. The pycdec module supports the cdec decoder. It enables Python coders to use cdec's fast C++ implementation of core finite-state and context-free inference algorithms for decoding and alignment. Its high-level interface allows developers to build integrated MT applications that take advantage of the rich Python ecosystem without sacrificing computational performance. The modular architecture of pycdec separates search space construction, rescoring and inference.

The cdec module includes implementations of the basic evaluation metrics (BLEU, TER and CER), exposed in Python via the cdec.score module. For a given (reference-hypomonograph) pair, sufficient statistics vectors

(SufficientStats) can be computed. These vectors are then summed for all sentences in the corpus, and the final result is converted into a real-valued score.

Before aligning a large data file, it is important to determine the proper comparators and acceptance rates for each comparison heuristic. Files of 1,000–10,000 lines result in the best performance. It is recommended to first evaluate each comparison method separately, and then combine the best ones in a specific scenario. For this purpose, using a binary search method to determine the best threshold factor value is recommended.

4.7.3 Filtering results

Experiments were performed to compare the performance of the proposed method with human judgment. First, the Polish side of corpora was repaired, spell-checked and cleaned by human translators (collegues and students), who helped accomplish this task on a volunteer basis. They were supposed to remove lines with translation into other languages or improper translations. The same task was performed with the filtering tool. Table 21 shows how many sentence pairs remained after the filtering, as well as the number of distinct words and their forms in the TED data.

Table 21. Filtering performance on TED data.

	Sentence Pairs	PL Vocabulary	EN Vocabulary
Baseline	185,637	247,409	118,160
Human	181,493	218,426	104,177
Filtering Tool	151,288	215,195	104,006

Table 22. SMT results.

	DATA SET	BLEU	NIST	TER	METEOR
Baseline	2011	16.69	5.20	67.62	49.35
	2012	19.76	5.66	63.50	52.76
	2013	16.92	5.30	65.90	49.37
	AVG	**17.79**	**5.38**	**65.67**	**50.49**
Human	2011	17.00	5.33	66.38	49.90
	2012	20.34	5.78	62.18	53.55
	2013	17.51	5.39	64.88	50.14
	AVG	**18.28**	**5.50**	**64.48**	**51.19**
Filtering Tool	2011	16.81	5.22	67.21	49.43
	2012	20.08	5.72	62.96	53.13
	2013	16.98	5.33	65.66	49.70
	AVG	**17.94**	**5.42**	**65.27**	**50.75**

Finally, the SMT systems were trained on the original and cleaned data to show the data's influence on the results. The results were compared using the BLEU, NIST, METEOR and TER metrics. The results are shown in Table 22. In these experiments, tuning was disabled because of the known MERT instability [45]. Test and development sets were taken from the IWSLT evaluation campaigns (from different years) and cleaned before usage with the help of human translators.

The significance of the results of Table 24 was calculated. The baseline was compared to human work and to the filtering tool. In addition, human work was compared with the proposed filtering tool. The calculation was done using the two-tailed Wilcoxon test. The results presented in Table 23 show that the differences among the results were highly significant, that the filtering tool performed better than the baseline method, and that humans performed better than both.

Table 23. Significance results for Table 21.

	Baseline vs. Human	Baseline vs. Filtering	Human vs. Filtering
P-value (Two-tailed):	0.000***	0.000***	0.003***

SMT experiments were conducted using the baseline system settings. For this evaluation, randomly selected test sets were chosen. Each contained 1,000 sentences. Lastly, the average values of metrics were compared.

For Wikipedia comparable corpus filtration, an initial experiment based on 1,000 randomly selected bi-sentences from the corpora was conducted. The filtering tool processed the data. Most of the noisy data was removed, but some good translations were lost in the process. Nevertheless, the results are promising, and it is intended to filter entire corpora in the future. It also must be noted that the filtering tool was not adjusted to this specific text domain. The results are presented in Table 24.

Table 24. Initial Wikipedia filtering results.

Number of sentences in base corpus	1,000
Number of poor sentences in test corpus	182
Number of poor filtered sentences	154
Number of good filtered sentences	12

In general, it is a very important to create high-quality parallel text corpora. Obviously, employing humans for this task is far too expensive because of the need for tremendous amounts of data to be processed. Analysis of the results of these experiments leads to the conclusion that the solution fits somewhere between a noisy parallel corpus and a corpus improved by human translators.

The proposed approach is also language independent for languages with similar structure to PL or EN. The results show that the proposed method performed well in the statistical machine translation evaluation. The method can also be used for improvement of noisy data, as well as comparable data. The filtering method provided a better score than the baseline system, and would most likely improve the output quality of other MT systems.

In summary, the bi-sentence extraction task is becoming more popular in unsupervised learning for numerous tasks. This method overcomes the disparities between English and Polish or any other West-Slavic language. It is a language-independent method that can easily be adjusted to a new environment, and it only requires the adjustment of initial parameters. The experiments show that the method performs well. The corpora used here increased the MT quality in a wide text domain, the TED Talks. It can be assumed that even small differences can make a positive influence on real-life translation scenarios. From a practical point of view, the method requires neither expensive training nor language-specific grammatical resources, while producing satisfactory results.

4.7.4 Alignment evaluation results

Experiments were performed in order to compare the performance of the proposed method with several other sentence alignment implementations on the TED lectures.

The additional aligners, all created to develop parallel corpora, used in this experiment were: Bleualign, Hunalign, ABBYY Aligner, Wordfast Aligner,[25] and Unitex Aligner. Table 25 shows the results. The performance of the aligners was scored using a sentence alignment metric prepared specially for this purpose. The input data for this experiments were 100 transcriptions from TED lectures.

Table 25. Experimental results for sentence alignment.

Aligner	Score
Proposed Method	98.94
Bleualign	96.89
Hunalign	97.85
ABBYY Aligner	84.00
Wordfast Aligner	81.25
Unitex Aligner	80.65

---

[25] https://www.wordfast.net/wiki/Wordfast_Aligner

The special metric was needed to evaluate sentences properly aligned but built from synonyms or with a different phrase order. The following weights were chosen by empirical research. For an aligned sentence, 1 point is given. For a misaligned sentence, a –0.2 point penalty is given. For web service translations, 0.4 points are given. For translations due to disproportion between input files, 1 point is given (when one of two files included more sentences). The score is normalized to fit between 1 and 100. A higher value is better. A floor function can be used to round the score to an integer value. Point weights were determined by empirical research and can be easily adjusted if needed. The score S is defined as:

$$S = floor(\frac{20(5A - M + 2T + 5|D|)}{L}) \qquad (77)$$

where A is the number of aligned sentences, M is the number of misaligned sentences, T is the number of translated sentences, D is the number of lines not found in both language files (one file can contain some sentences that do not exist in the other one) and L is the total number of output lines.

Clearly, the first three aligners scored well. The proposed method is fully automatic. It is important to note that Bleualign does not translate text and requires that it be done manually.

As discussed earlier, it is important not to lose lines of text in the alignment process. Table 26 shows the total lines resulting from the application of each alignment method. Five TED transcriptions were randomly selected for this part of the experiment.

Table 26. Experimental results.

Aligner	Lines
Human Translation	1,005
Proposed Method	1,005
Bleualign	974
Hunalign	982
ABBYY Aligner	866
Wordfast Aligner	843
Unitex Aligner	838

Almost all the aligners compared, other than the proposed method, lose lines of text as compared to a reference human translation. The proposed method lost no lines.

Table 27. Human vs. automatic alignment comparison results.

Aligner	BLEU	NIST	MET	TER	% of correctness
Human Translation	100	15	100	0	100
Proposed Method	98.91	13.81	99.11	1.38	98
Bleualign	91.62	13.84	95.27	9.19	92
Hunalign	93.10	14.10	96.93	6.68	94
ABBYY Aligner	79.48	12.13	90.14	20.01	83
Wordfast Aligner	85.64	12.81	93.31	14.33	88
Unitex Aligner	82.20	12.20	92.72	16.39	86

For the purpose of showing the output quality with an independent metric, it was decided to compare results with BLEU, NIST, METEOR and TER (the lower the better), in a comparison with human-aligned texts. Those results are presented in Table 27.

In general, sentence alignment algorithms are very important in creating parallel corpora. Most aligners are not fully automatic, but the one proposed here is, which gives it a distinct advantage. It also enables creation of a corpus when sentences exist in only a single language. The proposed approach is also language independent for ones with similar structure to PL or EN.

The results show that the proposed method performed very well in terms of the metric. It also lost no lines of text, unlike the other aligners. This is critical to the end goal of obtaining a translated text. The proposed alignment method also scored better when compared to typical machine translation metrics, and will most likely improve MT system output quality.

4.8 Baseline system training

A procedure involving a number of steps was used to train the baseline systems in all the experiments in this research. Processing of the corpora was accomplished, including tokenization, cleaning, factorization, conversion to lower case, splitting and a final cleaning after splitting. The baseline maximum sentence length was set to 80 words, and phrase length was set to 5 words. Training data was processed, and a language model was developed. Tuning was performed for each experiment. Lastly, the experiments were conducted.

The baseline system testing was done using the Moses open source SMT toolkit with its Experiment Management System (EMS) [8]. The SRI Language Modeling Toolkit (SRILM) [9] with an interpolated version of the Kneser-Ney discounting (interpolate –unk –kndiscount) was used for 5-gram language model training. The MGIZA++ tool was used for word and phrase alignment. KenLM [10] was used to binarize (transform features of a text entity into vectors of numbers) the language model, with the lexical reordering set to the msd-bidirectional-fe model [11]. The symmetrization method was set to grow-diag-final-and for word alignment processing [8], as discussed in Section 4.1.2.

4.9 Description of experiments

This section describes the experiments that were designed for the present research. First, the text alignment processing used in the experiment is discussed. Experimental results will be presented and discussed in Chapter 5.

4.9.1 Text alignment processing

For the experiments, the GIZA++ tool was used to create word alignments. GIZA++ is an SMT toolkit that can be used in training IBM Models 1–5 and also a hidden Markov model (HMM) alignment model. The default package of this toolkit also contains the mkcls tool, which can generate the word classes necessary for training some of the alignment models [116]. The mkcls tool trains word classes based on maximum likelihood criterion. The resulting word classes are especially suited for language models or statistical translation models [117, 118]. The GIZA++ itself is an extension of the well-known program GIZA, which was developed by the SMT team at the Center for Language and Speech Processing at Johns-Hopkins University (CLSP/JHU) during a summer workshop in 1999. It includes many additional features, the most important of which are IBM models 4 and 5, alignment models depending on word classes, and the HMM alignment model (Baum-Welch training, Forward-Backward algorithm, empty word, dependency on word classes and transfer to fertility models). It also includes a variant of IBM models 3 and 4, which enable training of the parameter p0 (probability of translation into an empty word), various smoothing techniques for fertility, and distortion/alignment parameters [116]. To be more precise, in this research the MGIZA variant of GIZA++ is used. MGIZA extends GIZA++ with multi-threading support, resumption of training, and the ability to conduct incremental training [11].

During word alignment, many different heuristics can be used. By default, the symmetrization heuristic grow-diag-final-and was used it this research. First, this heuristic intersects two-way direction alignments obtained from GIZA++, so that only the alignment points that occurred in both alignments remained. In the second phase, additional alignment points existing in their union were added. The growing step adds potential alignment points of unaligned words and neighbors. Neighborhood can be set directly to left, · right, top, bottom or diagonal (grow-diag). In the final step, alignment points between words from which at least one is unaligned are added (grow-diag-final). If the grow-diag-final-and method is used, an alignment point between two unaligned words appears. To illustrate this heuristic, see the example in Figure 31 with the intersection of the two alignments for the two sentences [8].

Second, additional alignment points that lie in the union of the two alignments are added, as shown in Figure 32.

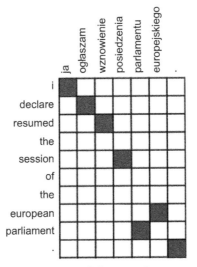

Figure 31. Intersection of alignment of two sentences.

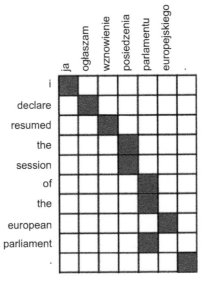

Figure 32. Union of two alignments.

This arrangement has a glaring slip—the alignment of the two verbs is mixed up: "resumed" is aligned to "wznowienie," and "adjourned" is connected to "odroczonego"; however, it ought to be the other way around.

Within the scope of this monograph, the usage of alternative tools to GIZA, such as Berkeley Aligner and Dyer's Fast Align, and different phrase models (Hierarchical or Target Syntax) was investigated. Factored training with stemmed word alignment was also examined [85].

4.9.2 *Machine translation experiments*

This section describes the design of the machine translation experiments performed during this research. These experiments involved the machine translation of TED lectures, subtitles, EuroParl proceedings and medical texts, as well as pruning experiments.

4.9.2.1 *TED lectures translation*

Experiments conducted on the TED lectures were motivated by the IWSLT Evaluation Campaigns [154], for which machine speech translation systems were prepared. Unfortunately, there is not much focus on Polish in the campaign, so there is almost no additional data in Polish in comparison to a huge amount of data in, for example, French or German. To remedy this disproportion, additional data had to be collected. The perplexity metric, defined in Section 2.4.3, was used to determine the data that was obtained during this research.

Some of the data used in the experiments was collected from the OPUS [14] project page, some from other small projects, and the rest manually using web crawlers. The following corpora were created:

- A Polish-English dictionary (bilingual parallel)
- Additional (newer) TED Talks data sets not included in the original training, development or test data (bilingual parallel)
- E-books (monolingual PL + monolingual EN)
- Proceedings of UK House of Lords (monolingual EN)
- Subtitles for movies and TV series (monolingual PL)
- Parliament and Senate proceedings (monolingual PL)
- Wikipedia comparable corpus (bilingual parallel)
- Euronews comparable corpus (bilingual parallel)
- Repository of the Polish-Japanese Academy of Information Technology diplomas (monolingual PL)
- PL monolingual web data crawled from main web portals like blogs, chip. pl, the Focus newsmonograph archive, interia.pl, wp.pl, onet.pl, money.

pl, Usenet, Termedia, Wordpress web pages, Wprost newsmonograph archive, Wyborcza newsmonograph archive, Newsweek newsmonograph archive, etc.

The experiments used many very small models merged together and included both unsmoothed data and data smoothed with the Kneser-Ney algorithm. The MIT language modeling (MITLM) [55] toolkit was used for evaluation. An evaluation set the of dev2010 data, which was prepared for tuning during IWSLT, was used. Its dictionary includes 2,861 words.

Perplexity was also examined for smoothed and unsmoothed data from EMEA texts, a Graphical User Interface (GUI) localization file, the European Central Bank (ECB) corpus, the OpenSubtitles [14] corpus of movies and TV series subtitles, a web crawl of the euronews.com web page (EUNEWS), data from bookshop.europa.eu and additional TED data. Parallel corpora were also extracted from Wikipedia and EUNEWS comparable corpora. Perplexity would determine if improvements could be expected from the data alone. It should be emphasized that both automatic and manual preprocessing of this training information was required.

4.9.2.1.1 Word stems and SVO word order

As previously described, there is a disproportionate vocabulary size between the PL and EN languages. This problem remained in the TED data, with 41,163 English words and 92,135 Polish words. Gerber and Yang [213] indicated that vocabulary size is critical to high-quality translations. For example, in the TED data, the mismatch between vocabulary sizes makes it extremely difficult to determine which Polish word best corresponds to a source English word. A study by Birch et al. [214] indicated that the vocabulary size of a target language is an indicator of morphological complexity and has a significant effect on translation performance. In addition, large differences in morphology between two languages may be even more relevant to translation performance than their complexity, using Finnish as an example [214]. So, disproportionate vocabulary sizes pose a significant problem for translation.

One of the solutions to the problem of disproportionate vocabularies, according to Bojar [58], is to use stems instead of surface forms. Applying this solution reduced the Polish vocabulary size to 40,346 words. Such a solution also requires creation of an SMT system from Polish stems to plain Polish. Subsequently, morphosyntactic tagging, using the Wroclaw Natural Language Processing (NLP) tools, was included as an additional information source for SMT system preparation. This can be also used as a first step for implementing a factored SMT system that, unlike a phrase-based system, includes morphological analysis, translation of lemmas and features, and the generation of surface forms. Incorporating additional linguistic information should improve translation performance [58].

As mentioned, stems extracted from Polish words are used instead of surface forms in order to overcome the problem of the huge difference in vocabulary sizes. Keeping in mind that in half of experiments the target language was English in the form of normal sentences, it was not necessary to introduce models for converting the stems to the appropriate grammatical forms. However, future work in translation into Polish could leverage this technique. For Polish stem extraction, a set of natural language processing tools available at http://nlp.pwr.wroc.pl was used. These tools can be used for:

- Tokenization
- Morphosyntactic analysis
- Shallow parsing for chunking
- Text transformation into feature vectors

The following two components were also used:

- MACA—a universal framework used to connect various morphological data
- WCRFT—this framework combines conditional random fields and tiered tagging

These tools used in sequence provide an XML output. The output includes the surface form of the tokens, stems and morphosyntactic tags. An example of such data is given below.

Wroclaw's tools[26] were used to tag morphosyntactic elements. Every tag in this tagset consists of specific grammatical classes with specific values for particular attributes. Furthermore, these grammatical classes include attributes with values that require additional specification. For example, nouns require specification of gender, number and case; adverbs require specification of an appropriate degree of an attribute. This causes segmentation of the input data, including tokenization of the words in a manner different from that of the Moses tool. On the other hand, this causes problems with building parallel corpora. This can be solved by placing markers at the end of input lines.

In the following example, where the pl.gen. "men" is derived from the sin. nom. "człowiek" (man) or pl.nom. "ludzie" (people), it can be demonstrated how one tag is used where, in the most difficult cases, more possible tags are provided.

---

[26] http://www.nlp.pwr.wroc.pl/en/tools-and-resources/narzedzia-przetwarzania-morfo-syntaktycznego

```
<tok>
<orth>ludzi</orth>
<lex disamb="1"> <base>człowiek</base>
<ctag>subst:pl:gen:m1</ctag></lex>
<lex disamb="1"> <base>ludzie</base>
<ctag>subst:pl:gen:m1</ctag></lex>
</tok>
```

In this example, only one form (the first stem) is used for further processing.

An XML extractor tool (in Python) was developed to generate three different corpora for the Polish language data:

• Word stems
• Subject-Verb-Object (SVO) word order
• Both the word stem and the SVO word order form

This enables experiments with these preprocessing techniques.

4.9.2.1.2 Lemmatization

Another approach to deal with disproportionate vocabulary sizes is using lemmas instead of surface forms to reduce the Polish vocabulary size, using PSI-TOOLKIT [59] to convert each Polish word into a lemma. This toolkit provides a tool chain for automatic processing of the Polish language and, to a lesser extent, other languages like English, German, French, Spanish and Russian (with a focus on machine translation). The tool chain includes segmentation, tokenization, lemmatization, shallow parsing, deep parsing, rule-based machine translation, statistical machine translation, automatic generation of inflected forms from lemma sequences and automatic post editing. The toolkit was used as an additional information source for SMT system preparation. It can also be used as a first step in implementing a factored SMT system that, unlike a phrase-based system, includes morphological analysis, translation of lemmas and features such as the generation of surface forms. Incorporating additional linguistic information should improve translation performance [60].

As previously mentioned, lemma extracted from Polish words are used instead of surface forms to overcome the problem of the huge difference in vocabulary sizes. For Polish lemma extraction, a tool chain that included tokenization and lemmatization from PSI-TOOLS was used [59].

These tools used in sequence provide a rich output that includes a lemma form of the tokens, prefixes, suffixes and morphosyntactic tags. Unfortunately, unknown words like names, abbreviations, or numbers, etc., are lost in the process. In addition, capitalization, as well as punctuation, is removed. To preserve this relevant information, a specialized tool was implemented based on differences between the input and output of PSI-TOOLS [59]. This tool restores most of the lost information.

4.9.2.1.3 Translation and translation parameter adaptation experiments

About a hundred experiments were conducted using test 2010 and development 2010 data to determine the best possible translation settings from Polish to English, and vice versa. During these experiments, the language model was extended with more data, and it was linearly interpolated. Other variations investigated included: Not using lower casing, changing maximum sentence length to 95 and setting the maximum phrase length to 6. In addition, experiments were conducted with the language model order set to 5 and 6, and with the Kneser-Ney and Witten-Bell discounting methods. During the training phase, lexicalized reordering methods of msd-bidirectional-fe and hier-mslr-bidirectional-fe were used. The system was also enriched by using the OSM [62]. The compound splitting feature was also used. Tuning was performed using the MERT tool with the batch-mira feature, and n-best list sizes of 100 and 150 were investigated. Lastly, all parallel data that were obtained were used. The data was adapted using Modified Moore-Lewis filtering [60].

Because of a much larger dictionary, the translation from EN to PL is significantly more complicated. Similar to PL-EN translation, variations in the use of lower casing, a maximum sentence length of 85, and a maximum phrase length of 7 were investigated. In addition, language model orders of 5 and 6, and the Kneser-Ney and Witten-Bell discounting methods were examined. During the training phase, lexicalized reordering methods of msd-bidirectional-fe and tgttosrc were used. The system was also enriched by using the OSM. The compound splitting feature was used, and punctuation normalization was performed. Tuning was performed using the MERT tool with the batch-mira feature, and n-best list sizes of 100 and 150 were examined, as well as training of a hierarchical phrase-based translation model [61].

All available parallel data was used. The data was adapted using Modified Moore-Lewis filtering [60]. Finally, the IWSLT [154] Polish-to-English and English-to-Polish translation evaluation experiments were performed on the best systems with test data from 2010–2014.

4.9.2.2 *Subtitles and EuroParl translation*

Experiments were designed for the movie subtitles domain and for European Parliament Proceedings based on phrase-based and factored systems enriched with Part of Speech (POS) tags. The use of compound splitting and true casing was optional. Language models were chosen, based on perplexity measure, and linearly interpolated [3].

4.9.2.3 *Medical texts translation*

A number of experiments were conducted using test and development data randomly selected and removed from medical texts from the EMEA corpora. An implemented script generated 1,000 segments for Polish-to-English and English-to-Polish translations. The experiments were measured by the BLEU, NIST, TER, RIBES and METEOR metrics.

Experiment 00 evaluated the baseline system. Each of the subsequent experiments involved a separate modification to the baseline. Experiment 01 applied truecasing and punctuation normalization. Experiment 02 was enriched by using the OSM described in Chapter 2 [71].

Experiment 03 was based on a factored model that allows additional annotation at the word level, which may be exploited in various models. Here, part-of-speech tagged English data was used as the basis for the factored phrase model [72]. Hierarchical phrase-based translation combines the strengths of phrase-based and syntax-based translation. It uses phrases (segments or blocks of words) as units for translation and uses synchronous context-free grammars as rules (syntax-based translation). Hierarchical phrase models allow for rules with gaps. Since these are represented by non-terminals and such rules are best processed with a search algorithm that is similar to syntactic chart parsing, such models fall into the class of tree-based or grammar-based models. Such a model was used in Experiment 04. The target syntax model implies the use of linguistic annotation for non-terminals in hierarchical models. This requires running a syntactic parser. For this purpose, the "Collins" [73] statistical natural language parser was used in Experiment 05.

Experiment 06 was conducted using stemmed word alignment. The factored translation model training makes it very easy to set up word alignment based on word properties other than the surface forms of words. One relatively popular method is to use stemmed words for word alignment. There are two main reasons for this. For morphologically-rich languages, stemming overcomes the data sparsity problem. Second, GIZA++ may have difficulties with very large vocabulary sizes, and stemming reduces the number of unique words.

Experiment 07 used Dyer's Fast Align [74], which is another alternative to GIZA++. It runs much faster and often gives better results, especially for language pairs that do not require large-scale reordering. In Experiment 08, settings recommended by Koehn in his SMT system from WMT'13 were used [75]. In Experiment 09, the language model discounting was changed to Witten-Bell [76]. Lexical reordering was set to hier-mslr-bidirectional-fe in Experiment 10. In Experiment 11, the compound splitting feature was used. Lastly, for Experiment 12, the settings recommended in [18] were used.

4.9.2.4 Pruning experiments

Phrase tables often grow enormously large, because they contain a lot of noisy data and store many very unlikely hypotheses. This is problematic when a real-time translation system that requires loading into memory is required. Quirk and Menezes in [123] argue that extracting only minimal phrases (the smallest phrase alignments, each of which map to an entire sentence pair) will negatively affect translation quality. This monograph is also the basis of the n-gram translation model [124, 125]. Discarding unlikely phrase pairs based on their significance tests drastically reduces phrase table size and may even yield increases in performance, according to [126]. Wu and Wang [127] propose a method for filtering the noise in the phrase translation table based on a logarithmic likelihood ratio. On the other hand, Kutsumi et al. [128] use a support vector machine for cleaning phrase tables. Eck et al. [129, 130] suggest pruning the translation table based on how often a phrase pair occurred during the decoding step and how often it was used in the best translation. The Moses-based method introduced by Johnson et al. [126] was implemented and used in the experiments.

Absolute and relative filtration in these experiments were based on in-domain dictionary filtration performed using different rules, for comparison. One rule (absolute) was to retain a sentence if at least a minimum number of words from a dictionary appeared in it. A second rule (relative) was to retain a sentence if at least a minimum percentage words in the sentence from a dictionary appeared in it. A third rule was to retain only a set number of the most probable translations. Finally, an experiment was performed using a combination of pruning and absolute pre-filtering.

4.9.3 Evaluation of obtained comparable corpora

This section describes the evaluation of the obtained comparable corpora.

4.9.3.1 Native Yalign method

To evaluate the corpora using the native Yalign method, each corpus was divided into 200 segments, and 10 sentences were randomly selected from each segment. This methodology ensured that the test sets covered the entire corpus. The selected sentences were removed from the corpora. The testing system was trained with the baseline settings. In addition, a system was trained with extended data from the Wikipedia corpora. Lastly, Modified Moore-Lewis filtering was used for domain adaptation of the Wikipedia corpora. The monolingual part of the corpora was used as a language model and was adapted for each corpus by using linear interpolation [82].

An evaluation was conducted using test sets built from 2,000 randomly-selected bi-sentences taken from each domain. For scoring purposes, four well-known metrics that show high correlation with human judgments were used: BLEU, the NIST metric, METEOR and TER.

Starting from the baseline system tests in the PL-to-EN and EN-to-PL directions, the effects of the following changes were investigated: Extending the language model, interpolating it, supplementing the corpora with additional data and filtering additional data with Modified Moore-Lewis filtering [82]. It should be noted the language models were extended after MML filtration.

4.9.3.2 Improved Yalign method

Experiments were designed and performed in order to evaluate methods of improving the performance of Yalign. Several improvements were developed. First, speed improvements were made by introducing multi-threading to the algorithm, using a database instead of plain text files or Internet links, and using GPU acceleration in sequence comparison. More importantly, two improvements were made to the quality and quantity of the mined data. The A* search algorithm was modified to use Needleman-Wunsch, and a tuning script of mining parameters was developed. The TED corpus was used to demonstrate the impact of the improvements (This was the only classifier used in mining phase.). The data mining approaches used for the experiments were: Directional (PL->EN) mining, bi-directional (PL->EN and EN->PL) mining, bi-directional mining with Yalign using a GPU-accelerated version of the Needleman-Wunsch algorithm and mining using a Needleman-Wunsch version of Yalign that was tuned.

The computation time and quality of the data obtained by these various methods for SMT was also investigated using two additional experiments. First, a speed comparison was made using the different Yalign-based algorithms.

Next, the quality of the data resulting from the different methods was investigated. A total of 1,000 comparable articles were randomly selected from Wikipedia and aligned using each of the methods. MT experiments were then conducted in order to verify potential gains in translation quality on the data that was tuned and aligned using a different heuristic. The TED, EUP, EMEA and OPEN domains were used for this purpose. For each of the domains, the system was trained using baseline settings. The additional corpora were used in the experiments by adding parallel data to the training set with Modified Moore-Lewis filtering and by adding a monolingual language model with linear interpolation.

To verify the results, an SMT system was trained using only data extracted from comparable corpora (not using the original in domain data). The mined data was also used as a language model. The evaluation was conducted using the same test sets as before.

4.9.3.3 Parallel data mining using tool pipeline

In this chapter, a chain of tools was described as a means for mining parallel data. To evaluate the quality and quantity of parallel data automatically extracted from comparable corpora, 20 bilingual documents were randomly selected from Wikipedia. Some of them differed greatly in vocabulary, amount of text, and parallelism. Human translators were asked to manually align the articles on the sentence level. An experiment was designed in order to compare the performance of the human translators, Hunalign, and the author's new method for obtaining, mining, and filtering very parallel, noisy-parallel and comparable corpora. The comparison was designed to assess how many sentences were correctly aligned and how many had alignment errors. It is important to note that the author's method is independent of language, can easily be adjusted to a new environment and requires only parallel corpora for initial training.

4.9.3.4 Analogy-based method

To evaluate the analogy-based method for obtaining comparable corpora, the TED corpus was divided into 200 segments, and 10 sentences were randomly selected from each segment. This methodology ensured that the test sets covered the entire corpus. The selected sentences were removed from the corpora. The baseline system was trained, as well as a system with extended training data obtained from the Wikipedia corpus using the analogy-based method. Lastly, Modified Moore-Lewis filtering was employed for domain adaptation of the analogy-based corpus. Additionally, the monolingual part of the analogy corpora was used as a language model and adapted using linear interpolation [10].

5

Results and Conclusions

In this chapter, the results of the machine translation experiments will be presented. The alignment of texts and adaptation of translation parameters for the needs of Polish-English translation will be discussed. Results of comparable corpora mining for parallel sentences will be provided and evaluated by improvements in machine translation systems. Additionally, bilingual and monolingual data that was obtained during this research is described.

5.1 Machine translation results

This section describes the results and improvements obtained for the statistical machine translation system evaluated on the TED lectures [15], European Parliament Proceedings [36], Movie Subtitles [14] and European Medicines Agency [35] medical documents text domains. Additionally, experiments on pruning of phrase tables to reduce memory consumption were performed. The TED experiments are also divided into experiments on lemmatization with PSI-TOOLS [59], conversion to word stems using the Wroclaw NLP tools[27] and experiments that deal with modification of word order within the sentences.

5.1.1 TED lectures experiments

The TED Talks do not have any specific domain. Statistical Machine Translation by definition works best when very specific domain data is used.

---

[27] www.nlp.pwr.wroc.pl

The TED data is a mix of various, unrelated topics. This is most likely the reason why we cannot expect big improvements with this data and generally low scores in translation quality metrics.

There is almost no additional data in Polish in comparison to a huge amount of data in, for example, French or German. Some of the data were obtained from the OPUS [223] project page, some from another small projects and the rest was collected manually using web crawlers.

The new data were:

- A Polish—English dictionary (bilingual parallel)
- Additional (newer) TED Talks data sets not included in the original train data (we crawled bilingual data and created a corpora from it) (bilingual parallel)
- E-books (monolingual PL + monolingual EN)
- Proceedings of UK House of Lords (monolingual EN)
- Subtitles for movies and TV series (monolingual PL)
- Parliament and senate proceedings (monolingual PL)
- Wikipedia Comparable Corpus (bilingual parallel)
- Euronews Comparable Corpus (bilingual parallel)
- Repository of PJIIT's diplomas (monolingual PL)
- Many PL monolingual data web crawled from main web portals like blogs, chip.pl, Focus newsmonograph archive, interia.pl, wp.pl, onet.pl, money.pl, Usenet, Termedia, Wordpress web pages, Wprost newsmonograph archive, Wyborcza newsmonograph archive, Newsweek newsmonograph archive, etc.

"Other" in the table below stands for many very small models merged together. EMEA are texts from the European Medicines Agency, KDE4 is a localization file of that GUI, ECB stands for European Central Bank corpus, OpenSubtitles [12] are movies and TV series subtitles, EUNEWS is a web crawl of the euronews.com web page and EUBOOKSHOP comes from bookshop.europa.eu. Lastly, bilingual TEDDL is additional TED data.

Data perplexity was examined by experiments with the TED lectures. Perplexities for the dev2010 data sets are shown in Table 28. In the table, "PPL" indicates perplexity values without smoothing of the data. "PPL + KN" indicates perplexity values with Kneser-Ney smoothing of the data.

Table 28. Data perplexities for dev2010 data set.

Data set	Dictionary	PPL	PPL + KN
Baseline train.en	44,052	221	223
EMEA	30,204	1,738	1,848
KDE4	34,442	890	919
ECB	17,121	837	889
OpenSubtitles	343,468	388	415
EBOOKS	528,712	405	417
EUNEWS	21,813	430	435
NEWS COMM	62,937	418	465
EUBOOKSHOP	167,811	921	950
UN TEXTS	175,007	681	714
UK LORDS	215,106	621	644
NEWS 2010	279,039	356	377
GIGAWORD	287,096	582	610
DICTIONARY	39,214	8,629	8,824
OTHER	13,576	492	499
WIKIPEDIA	682,276	9,131	9,205
NEWSMONOGRAPHS	608,186	10,066	10,083
WEB PORTALS	510,240	731	746
BLOGS	76,697	3,481	3,524
USENET	733,619	8,019	8,034
DIPLOMAS	353,730	32,345	32,58
TEDDL	47,015	277	277

5.1.1.1 Word stems and SVO word order

Results of the experiments on the effect of stemming and SVO word order, conducted with the use of the test data from 2010–2013, are shown in Tables 29 and 30 for the Polish-to-English and English-to-Polish translation, respectively. Procedure taken to convert Polish part of the corpora was described in the Section 4.9.2.1.1. Python tool was implemented and utilized for this conversion using annotated data. They are measured by the BLEU, NIST, TER and METEOR metrics.

Note that a lower value of the TER metric is better, while the other metrics are better when their values are higher. BASE stands for baseline system with no improvements. COR is a system with corrected spelling in Polish data (automatically, with the usage of filtering tool described in Section 4.7.2 and a cleaning tool programmed for this purpose). STEM is a system using stems in

Table 29. Polish-to-English translation.

System	Year	BLEU	NIST	TER	METEOR
BASE	2010	16.02	5.28	66.49	49.19
COR	2010	16.09	5.22	67.32	49.09
BEST	2010	20.88	5.70	64.39	52.74
STEM	2010	13.22	4.74	70.26	46.30
SVO	2010	9.29	4.37	76.59	43.33
BASE	2011	18.86	5.75	62.70	52.72
COR	2011	19.18	5.72	63.14	52.88
BEST	2011	23.70	6.20	59.36	56.52
BASE	2012	15.83	5.26	66.48	48.60
COR	2012	15.86	5.32	66.22	49.00
BEST	2012	20.24	5.76	63.79	52.37
BASE	2013	16.55	5.37	65.54	49.99
COR	2013	16.98	5.44	65.40	50.39
BEST	2013	23.00	6.07	61.12	55.16
STEM	2013	12.40	4.75	70.38	46.36

Polish. SVO is a system with the subject-verb-object word order in a sentence. BEST is the system that produced the best results.

Table 30. English-to-Polish translation.

System	Year	BLEU	NIST	TER	METEOR
BASE	2010	8.49	3.70	76.39	31.73
COR	2010	9.39	3.96	74.31	33.06
BEST	2010	10.72	4.18	72.93	34.69
STEM	2010	9.11	4.46	74.28	37.31
SVO	2010	4.27	4.27	76.75	33.53
BASE	2011	10.77	4.14	71.72	35.17
COR	2011	10.74	4.14	71.70	35.19
BEST	2011	15.62	4.81	67.16	39.85
BASE	2012	8.71	3.70	78.46	32.50
COR	2012	8.72	3.70	78.57	32.48
BEST	2012	13.52	4.35	73.36	36.98
BASE	2013	9.35	3.69	78.13	32.52
COR	2013	9.35	3.70	78.10	32.54
BEST	2013	14.37	4.42	72.06	37.87
STEM	2013	13.30	4.83	70.50	35.83

Statistical significance tests were performed to evaluate how the improvements differ from each other. For each text domain, the significance of the obtained results were compared to each other. Changes with low significance were marked with "*", significant changes were marked with "**", and very significant with "***" in Tables 31 and 32.

Table 31. Significance of PL-EN experiments.

System	Year	P-value
COR	2010	0.1863
BEST	2010	0.0106**
STEM	2010	0.0014***
SVO	2010	0.0015***
COR	2011	0.8459
BEST	2011	0.0020***
COR	2012	0.0626*
BEST	2012	0.0032***
COR	2013	0.0227**
BEST	2013	0.0016***
STEM	2013	0.0001***

In the Polish-to-English experiments, the significance tests proved that the BEST and STEM systems were significantly different from BASE. COR was only significantly different during 2012 and 2013, and in 2012 it was only significant at the 10% level. For English-to-Polish translations, only COR and SVO did not significantly alter the baseline system. Other results were highly significant.

Table 32. Significance of EN-PL experiments.

System	Year	P-value
COR	2010	0.0059***
BEST	2010	0.0020***
STEM	2010	0.00859***
SVO	2010	0.8277
COR	2011	0.8433
BEST	2011	0.0000***
COR	2012	0.3344
BEST	2012	0.000***
COR	2013	0.1398
BEST	2013	0.0003***
STEM	2013	0.0127**

Several conclusions can be drawn from the experimental results presented here. Automatic and manual cleaning of the training files had some impact, among the variations examined, on improving translation performance, together with spelling correction of the data in Polish—although it resulted in better BLEU and METEOR scores, not always in better NIST or TER scores. In particular, automatic cleaning and conversion of verbs to stems improved translation performance in English-to-Polish translation, quite the contrary to Polish-to-English translation.

This is likely due to a reduction of the Polish vocabulary size. However, the improvement is too low to bother with training an intermediate translation system (from PL stems to PL lemmas). Changing the word order to SVO is quite interesting. It did not help at all in some cases, although one would expect it would. In the experiments from PL to EN, the score was always worse, which was not anticipated. On the other hand, in some of the EN to PL experiments, improvement could be seen. Although the BLEU score dramatically decreased and TER became slightly worse, NIST and METEOR showed better results than the baseline system.

In summary, converting Polish verbs to stems reduces the Polish vocabulary, which should improve English-to-Polish translation performance. Polish-to-English translation typically outscores English-to-Polish translation, even on the same data. This requires further evaluation.

5.1.1.2 Lemmatization

During the lemmatization experiments, the lemmatized version of the Polish training data was reduced to 36,065 unique words, and the Polish language model was reduced from 156,970 to 32,873 unique words. The results of the experiments are presented in Tables 33 and 34. Each experiment was performed only on the baseline data sets in the PL->EN and EN->PL directions. The year (official test sets from the IWSLT campaigns[28]) column shows the test set that was used in the experiment. If a year has the suffix "L," it means that it is a lemmatized version of the baseline system.

The experiments show that lemma translation to EN in each test set decreased the evaluation scores, while translation from EN to lemma for each set increased the translation quality. Such a solution also requires training of a system from lemma to PL to restore proper surface forms of the words. Such a system was trained, as well, evaluated on official tests sets from 2010–2014, and tuned on 2010 development data. The results for that system are presented

---

[28] www.iwslt.org

Table 33. PL Lemma to EN translation results.

YEAR	BLEU	NIST	TER	METEOR
2010	16.70	5.70	67.83	49.31
2010L	13.33	4.68	70.86	46.18
2011	20.40	5.71	62.99	53.13
2011L	16.21	5.11	67.16	49.64
2012	17.22	5.37	65.96	49.72
2012L	13.29	4.64	69.59	45.78
2013	18.16	5.44	65.50	50.73
2013L	14.81	4.88	68.96	47.98
2014	14.71	4.93	68.20	47.20
2014L	11.63	4.37	71.35	44.55

Table 34. EN to PL Lemma translation results.

YEAR	BLEU	NIST	TER	METEOR
2010	9.95	3.89	74.66	32.62
2010L	12.98	4.86	68.06	40.19
2011	12.56	4.37	70.13	36.23
2011L	16.36	5.40	62.96	44.86
2012	10.77	3.92	75.79	33.80
2012L	14.13	4.83	69.76	41.52
2013	10.96	3.91	75.95	33.85
2013L	15.21	5.02	68.17	42.58
2014	9.29	3.47	82.58	31.15
2014L	12.35	4.44	75.27	39.12

in Table 35. Even though the scores are relatively high, the results do not seem satisfactory enough to provide an overall improvement of the EN-LEMMA-PL pipeline over direct translation from EN to PL.

Table 35. Lemma to PL translation results.

YEAR	BLEU	NIST	TER	METEOR
2010	41.14	8.72	31.28	65.25
2011	41.68	8.68	30.64	65.99
2012	38.87	8.38	32.23	64.18
2013	40.27	8.30	31.67	64.44
2014	37.78	8.01	33.17	62.78

To confirm this prediction, an additional experiment, in which the English sentences were first translated into lemma and then the lemma translated into Polish surface forms, was conducted. The results of such a combined translation approach are shown in Table 36. They decrease the translation quality in comparison to direct translation from EN to PL.

Table 36. EN-> PL Lemma-> PL pipeline translation.

YEAR	BLEU	NIST	TER	METEOR
2010	7.47	3.45	76.17	29.16
2011	9.67	3.84	72.45	32.25
2012	8.26	3.39	78.40	29.60
2013	8.83	3.54	77.11	30.61
2014	6.98	3.10	83.81	27.71

Converting Polish surface forms of words to lemma reduces the Polish vocabulary, which should improve English-to-Polish and Polish-to-English translation performance. Polish-to-English translation typically outscores English-to-Polish translation, even on the same data. This was also expected in the experiments with lemma; nonetheless, the initial assumptions were not confirmed in empirical tests.

5.1.1.3 Translation results and translation parameter adaptation

These experiments were conducted using test 2010 and development 2010 data to determine the best possible settings for translation from Polish to English and from English to Polish. Starting from baseline system tests (BLEU: 16.70), the score was raised through extending the language model with more data and by interpolating it linearly. It was determined that not using lower casing, changing maximum sentence length to 95, and setting the maximum phrase length to 6 improves the BLEU score. The following settings and language models produced a BLEU score of 21.57: Language model order of 6, Witten-Bell discounting method, a lexicalized reordering method of hier-mslr-bidirectional-fe during training, enrichment through use of the OSM, compound splitting, tuning performed using the MERT tool with the batch-mira feature and an n-best list size of 150.

From the empirical experiments, it was concluded that the best results are obtained when sampling about 150,000 bi-sentences from in-domain corpora and by using filtering after word alignment. The ratio of data to be retained was set to 0.8, obtaining the best score of 23.74. The results are presented in Table 37.

For the baseline system, the BLEU score for EN-to-PL translation was 9.95. Similar to results in the PL-EN direction, it was determined that not using lower casing, changing maximum sentence length to 85 and setting the maximum phrase length to 7 improves the BLEU score. The best results were achieved using these settings and models: A language model order of 6, the Witten-Bell discounting method, a lexicalized reordering method of tgttosrc during training, system enrichment using the OSM, compound splitting, punctuation normalization, tuning using MERT with the batch-mira feature, an n-best list size of 150 and training using a hierarchical phrase-based translation model. These settings and language models produced a BLEU score of 19.81. All available parallel data was used. The data was adapted using Modified Moore-Lewis filtering.

From the empirical experiments, it was concluded that the best results are obtained when sampling about 150,000 bi-sentences from in-domain corpora and by using filtering after word alignment. The ratio of data to be retained was set to 0.9, producing the best score of 22.76. The results are presented in Table 37.

Table 37. Polish-to-English translation.

System	Year	BLEU	NIST	TER	METEOR
BASE	2010	16.70	5.70	67.83	49.31
BEST	2010	23.74	6.25	54.63	57.06
BASE	2011	20.40	5.71	62.99	53.13
BEST	2011	28.00	6.61	51.02	61.23
BASE	2012	17.22	5.37	65.96	49.72
BEST	2012	23.15	5.55	56.42	56.49
BASE	2013	18.16	5.44	65.50	50.73
BEST	2013	28.62	6.71	57.10	58.48
BASE	2014	14.71	4.93	68.20	47.20
BEST	2014	18.96	5.56	64.59	51.29

Results of the IWSLT evaluation experiments, performed on the best systems with test data from 2010–2014, are shown in Tables 37 and 38 for Polish-to-English and English-to-Polish translation, respectively. They are assessed using the BLEU, NIST, TER and METEOR metrics.

Several conclusions can be drawn from the experimental results presented here. Automatic and manual cleaning of the training files has a positive impact, among the variations of the experiments [18]. Obtaining and adapting additional bilingual and monolingual data produced the biggest effect on the translation quality itself. In each translation direction, using OSM [62] and adapting training and tuning parameters were necessary. The results could not be simply replicated from other experiments.

Table 38. English-to-Polish translation.

System	Year	BLEU	NIST	TER	METEOR
BASE	2010	9.95	3.89	74.66	32.62
BEST	2010	22.76	5.83	60.23	49.18
BASE	2011	12.56	4.37	70.13	36.23
BEST	2011	29.20	6.54	55.02	51.48
BASE	2012	10.77	3.92	75.79	33.80
BEST	2012	26.33	5.93	60.88	47.85
BASE	2013	10.96	3.91	75.95	33.85
BEST	2013	26.61	5.99	59.94	48.44
BASE	2014	9.29	3.47	82.58	31.15
BEST	2014	16.59	4.48	73.66	38.85

In the future, the author intends to try clustering the training data into word classes in order to obtain smoother distributions and better generalizations. Using class-based models has been shown to be useful when translating into morphologically-rich languages like Polish [79]. The author also wants to use unsupervised transliteration models [80], which have proven quite useful in MT for translating OOV words, for disambiguation and for translating closely-related languages [80]. This feature would most likely help us overcome the difference in vocabulary size, especially when translating into PL. Using a fill-up combination technique (instead of interpolation) may be useful when the relevance of the models is known *a priori*—typically, when one is trained on in-domain data and the others on out-of-domain data [57].

5.1.2 Subtitles and European Parliament proceedings translation results

Table 39 shows the results of the experiments on movie subtitles and European Parliament proceedings. The following abbreviations are used: E for EuroParl, and O for OpenSubtitles. If there is no additional suffix, it means that the test involved the baseline system trained on a phrase-based model. The suffix F (e.g., TF) means the factored model was used, T refers to data that was true-cased, and C means that a compound splitter was used. In the EuroParl experiments, the suffix L stands for the larger EN in-domain language model. Truecasing[29] and compound splitting[30] were performed for both languages using tools already implemented in the Moses toolkit. H stands for the highest

---

[29] http://www.statmt.org/moses/?n=Moses.SupportTools
[30] http://www.statmt.org/moses/?n=FactoredTraining.EMS

score that was obtained by combining methods and interpolating additional data. The results for movie subtitles are shown in Table 39. Table 40 contains results for the proceedings.

Table 39. Experiment results—Subtitles.

	PL-> EN				EN->PL			
	BLEU	*NIST*	*METEOR*	*TER*	*BLEU*	*NIST*	*METEOR*	*TER*
O	53.21	7.57	66.40	46.01	51.87	7.04	62.15	47.66
OC	53.13	7.58	66.80	45.70	–	–	–	–
OT	52.63	7.58	67.02	45.01	50.57	6.91	61.24	48.43
OF	53.51	7.61	66.58	45.70	52.01	6.97	62.06	48.22

Table 40. Experiment results—European Parliament.

	PL-> EN				EN->PL			
	BLEU	*NIST*	*METEOR*	*TER*	*BLEU*	*NIST*	*METEOR*	*TER*
E	73.18	11.79	87.65	22.03	67.71	11.07	80.37	25.69
EL	80.60	12.44	91.07	12.44	–	–	–	–
ELC	80.68	12.46	90.91	16.78	67.69	11.06	80.43	25.68
ELT	78.09	12.41	90.75	17.09	64.50	10.99	79.85	26.28
ELF	80.42	12.44	90.52	17.24	69.02	11.15	81.83	24.79
ELI	70.45	11.49	86.21	23.54	70.73	11.44	83.44	22.50
ELS	61.51	10.65	81.75	31.71	49.69	9.38	69.05	40.51
ELH	82.48	12.63	91.17	15.73	–	–	–	–

It was concluded from the results that compound splitting improved translation quality, but mostly in PL-EN translation. Factored training models also provide better translations, but improvement was gained mostly in EN-PL translation. The most likely reason is the more complex Polish grammar. Truecasing did not help as anticipated; in most experiments, scores were worse. Most likely, that data was already correctly cased. Tuning of training parameters for each set of data must be done separately (just like training higher-order models).

5.1.3 *Medical text translation experiments*

Tables 41 and 42 show the results of the medical text translation experiments that were performed using test and development data randomly selected and removed from the corpora.

Several conclusions can be drawn from the experimental results presented here. It was surprising that truecasing and punctuation normalization decreased the scores by a significant factor. Supposedly, the text was already properly cased and punctuated. In Experiment 02, it was observed that, quite strangely, OSM decreased some metrics results, whereas it usually increases translation quality. Similar results can be seen in the EN->PL experiments. Here, the BLEU score increased, but other metrics decreased.

Table 41. Polish-to-English translation of medical texts.

System	BLEU	NIST	METEOR	TER
00 – Baseline System	70.15	10.53	82.19	29.38
01 – Truecasing & Punct. Norm.	64.58	9.77	76.04	35.62
02 – Operation Sequence Model	71.04	10.61	82.54	28.33
03 – Factored Model	71.22	10.58	82.39	28.51
04 – Hierarchical Model	**76.34**	**10.99**	**85.17**	**24.77**
05 – Target Syntax Model	70.33	10.55	82.28	29.27
06 – Stemmed Word Alignment	71.43	10.60	82.89	28.73
07 – Dyer's Fast Align	71.91	10.76	83.63	26.60
08 – Koehn's settings from WMT'13	71.12	10.37	84.55	29.95
09 – Witten-Bell discounting	71.32	10.70	83.31	27.68
10 – Hierarchical lexical reordering	71.35	10.40	81.52	29.74
11 – Compound Splitting	70.34	10.64	82.65	28.22
12 – Wolk's settings from IWSLT'13	72.51	10.70	82.81	28.19

Table 42. English-to-Polish translation of medical texts.

System	BLEU	NIST	METEOR	TER
00 – Baseline System	69.18	10.14	79.21	30.39
01 – Truecasing & Punct. Norm.	61.15	9.19	71.91	39.45
02 – Operation Sequence Model	69.41	10.14	78.98	30.90
03 – Factored Model	68.45	10.06	78.63	31.62
04 – Hierarchical Model	**73.32**	**10.48**	**81.72**	**27.05**
05 – Target Syntax Model	69.21	10.15	79.26	30.88
06 – Stemmed Word Alignment	69.27	10.16	79.30	31.27
07 – Dyer's Fast Align	68.43	10.07	78.95	33.05
08 – Koehn's settings from WMT'13	67.61	9.87	77.82	29.95
09 – Witten-Bell discounting	68.98	10.11	78.90	31.13
10 – Hierarchical lexical reordering	68.67	10.02	78.55	31.92
11 – Compound Splitting	69.01	10.14	79.13	30.84
12 – Wolk's settings from IWSLT'13	67.47	9.89	77.65	33.32

Most of the other experiments worked as anticipated. Almost all of them raised the score a little bit or were confirmed with consistent scores for multiple metrics. The baseline system score was already very high, most likely because of the high formality of texts and repetitions at phrase level that occure across different leaflets. Unfortunately, Experiment 12, which was based on settings that provided the best system score on the IWSLT 2013 evaluation campaign, did not improve the quality of this data as much as it did previously. The most likely reason is that the data used in IWSLT did not come from a specific text domain, while here the SMT system dealt with a very narrow domain. It may also mean that training and tuning parameter adjustment may be required separately for each text domain if improvements cannot be simply replicated.

On the other hand, improvements obtained by training the hierarchical based model were surprising. In comparison to other experiments, Experiment 04 increased the BLEU score by the highest factor. The same, significant improvements can be observed in both the PL->EN and EN->PL translations, which most likely provide a very good starting point for future experiments.

Translation from EN to PL is more difficult. The results of these experiments are simply worse than for PL to EN. The most likely reasons for this, of course, are the complicated Polish grammar and the larger Polish vocabulary.

Applying machine translation to medical texts holds great promise to benefit patients, including travelers and others who do not speak the language of the country in which they need medical help. Improving access to the vast array of medical information on the Internet would be useful to patients, medical professionals and medical researchers.

Human interpreters with medical training are too rare. Machine translation could also help in communicating medical history, diagnoses, proposed medical treatments, general health information and the results of medical research. Mobile devices and web applications may be able to boost delivery of machine translation services for medical information.

5.1.4 Pruning experiments

In Table 43 experimental results are provided for use of the Moses-based pruning method, along with pre-processing of the data by filtering irrelevant sentences from it.

The absolute and relative filtration experimental results here use the following terminology:

Absolute N—Retain a sentence if at least N words from dictionary appear in it.

Relative N—Retain a sentence if at least N% of sentence is built by from dictionary.

Table 43. Pruning results.

Optimizer	Exp. No.	BLEU	Phrase Table (GB)	Reordering Table (GB)
None	**2**	**35.39**	**6.4**	**2.3**
Absolute 3	1	26.95	1.1	0.4
Absolute 2	7	30.53	2.5	0.9
Absolute 1	8	32.07	4.9	1.7
Relative 2.5	9	30.82	3.1	1.1
Relative 5	10	26.35	1.1	0.4
Relative 7.5	11	17.68	0.3	0.1
Pruning 30	**3**	**32.36**	**1.9**	**0.7**
Pruning 60	4	32.12	2.0	0.7
Pruning 90	5	32.11	2.0	0.75
Pruning 20	6	32.44	2.1	0.75
Absolute 1 + Pruning 20	**12**	**30.29**	**0.85**	**0.3**

Pruning N—Retain only the N most probable translations.

Experiment 12 used a combination of pruning and absolute pre-filtering.

Unfortunately, contrary to the referenced publications, the empirical experiments showed that pruning (at least for PL-EN pair) significantly decreases translation quality. Even though phrase and reordering tables are much smaller, the decrease in quality is questionable. What is interesting is that a substantial factor of quality lost to final phrase table size was regained by combining pre-filtering and pruning of the phrase table in Experiment 12.

5.1.5 MT complexity within West-Slavic languages group

When implementing Statistical Machine Translation systems (SMT) it is necessary to deal with many problems to achieve high quality translations. These problems include the need to align parallel texts in language pairs and clean parallel corpora to remove errors. This is especially true for real-world corpora developed from text harvested from the vast data available in the Internet. Out-Of-Vocabulary (OOV) words must also be handled, as they are inevitable in real-world texts [1]. The lack of sufficient parallel corpora is another significant challenge for SMT. Since the approach is statistical in nature, a significant amount of quality language pair data is needed in order to improve translation accuracy. In addition, very general translation systems that work in a general text domain have accuracy problems in specific domains. SMT systems are more accurate on corpora from a domain that is not too wide.

Another problem not yet addressed is adaptation of machine translation techniques for the needs of the specific language or event kind of a text. Such adaptation would be very beneficial especially for very diverse languages, such as Slavic-English. In this research we focus on very complex West-Slavic languages, which represent a serious challenge to any SMT system. They differ from other languages similar to English in grammar full complicated rules and elements, together with a big vocabulary (due to complex declension). The main reasons for its complexity are more cases, genders, animate and inanimate nouns, adjectives agreements with nouns in terms of gender, case, number and a lot of words borrowed from other languages which are often inflected similarly to those of West-Slavic origin. This greatly affects the data and data structure required for statistical models of translation. The lack of available and appropriate resources required for data input to SMT systems presents another problem, especially when vocabularies are disproportionate (which is the case).

In our research, we conduct experiments that try to overcome those disproportions and improve the baseline systems based on Moses SMT state of the art configuration. We conduct research on the Polish language as a representative candidate from West-Slavic group and try to translate it between English and Czech. In addition, we do comparison of translation between spoken and written texts because in recent research [2], the authors analysed the differences between spoken language based on lectures and written texts based on written news for machine translation. They showed that there are meaningful differences between those two text genres and that particular statistical modeling strategies should be independently applied to each translation task.

The article is structured as follows: Section 2 describes West-Slavic languages and main differences between Polish and English languages, Section 3 contains comparison of spoken and written texts used in this research. Section 4 describes our machine translation systems and evaluation methods. Lastly, we draw conclusions in the Section 5.

5.1.5.1 West-Slavic languages group

West-Slavic languages are part of Slavic languages (alongside East- and South-Slavic), which are used by approximately 56 million people in the Central Europe. West-Slavic languages are divided into: Lechitic languages (Polish, Silesian and Kashubian, the latter two being dialects of Polish), Sorbian languages (Upper and Lower) and Czech-Slovak (Czech and Slovak) [3].

Approximately, a native speaker of Polish actively uses not much more than 12,000 words in their everyday language, whereas passively, they will use about 30,000 words. However, research of philologist from Bialystok University shows that knowledge of 1,200 most commonly used Polish words is enough to communicate effectively [4].

Kashubian, is considered either a separate language or a dialect of Polish. Often, indirect position is adopted, and terms such as "language" or "dialect" are avoided. Instead, Kashubian is classified as an "ethnolect". In law, Kashubian is regarded as a regional language. Every day, around 108,000 Polish people use Kashubian [5].

Silesian ethnolect is a set of Silesian jargons, possibly mixing into multiple dialects. It is used by indigenous inhabitants of Upper Silesia and a small population of Lower Silesia. We know that languages like literary Polish, Czech (particularly from Moravian, formerly functioning as a separate language), German (mostly Germanic dialect of Silesia) and Slovak played a huge role in the process of formation of the Silesian ethnolect, as we can trace many loanwords from those languages [6].

The Sorbian languages is a group consisting of two closely related West Slavic languages: Sorbian and Lower Sorbian. They are used by the Sorbs (totalling approximately 150,000 people, most of whom speak only German) in Lausitz in eastern Germany. Both languages are endangered Sorbian languages (especially Lower Sorbian; the number of active users is estimated at several thousand; mostly used only by the older generation, while the Upper Lusatian's users total approximately 55,000 people). Lower Sorbian shares more similarities with Polish, while Sorbian with Czech and Slovak.

Czech derives from Proto-Indo-European through the Proto-Slavic. It developed orally in the tenth century and its first artefacts date back to the thirteenth century. Late Middle Ages was a period of flourishing Czech and its strong influence on other languages (including Polish).

Silesian and Kashubian are very similar to Polish in terms of grammar. Both Upper and Lower Sorbian have the dual for nouns, pronouns, adjectives and verbs; very few known living Indo-European languages retain this feature as a productive aspect of the grammar. For example, the word ruka is used for one hand, ruce for two hands, and ruki for more than two hands [6].

There are four grammatical genders in Slovak language: Animate masculine, inanimate masculine, feminine and neutral. In popular description, the first two genders are often covered under common masculine gender. There is the singular and the plural numbers. The following morphological cases are present in Slovak: The nominative case answers the question Who / What; the genitive case (of whom); the dative case (to whom); the accusative case (whom); the locative case (used after the prepositions); the instrumental case (by means of whom).

An adjective's, pronoun's and, to some extent, also the number's person, gender and case decline according to the noun. An adjective always precedes the corresponding noun. The comparative is formed by replacing the adjective ending –ý/y/i/í by –ejší or –ší. There are exact rules for the choice between these two endings and there are several irregular comparatives. The superlative is formed as follows: naj+comparative. The comparative and superlative of adverbs (which end in –o, –e or –y in the basic form) are formed by simply replacing the –(ej)ší from the adjective by –(ej)šie. The verb (predicate) agrees in person and number with its subject [4].

Czech is very similar to Polish and Slovak. The most recognizable thing in Czech, which occurs also in Polish, is the fact that one word is often sufficient to express what English can only achieve by using multiple words.

5.1.5.2 Differences between Polish and English languages

As already discussed in detail in the introduction, Polish and English differ in syntax and grammar. English is a positional language, which means that the syntactic order (the order of words in a sentence) plays a very important role, particularly due to the limited inflection of words (e.g., lack of declension endings). Sometimes, the position of a word in a sentence is the only indicator of the sentence's meaning. In a Polish sentence, a thought can be expressed using several different word orderings, which is not possible in English. For example, the sentence "I bought myself a new car" can be written in Polish as "Kupiłem sobie nowy samochód", or "Nowy samochód sobie kupiłem", or "Sobie kupiłem nowy samochód" or "Samochód nowy sobie kupiłem." The only exception is when the subject and the object are in the same clause and the context is the only indication which is the object and which is subject. For example, "Mysz liże kość (A mouse is licking a bone.)" and "Kość liże mysz (A bone is licking a mouse.)".

Differences in potential sentence word order make the translation process more complex, especially when using a phrase-model with no additional lexical information [7]. In addition, in Polish it is not necessary to use the operator, because the Polish form of a verb always contains information about the subject of a sentence. For example, the sentence "On jutro jedzie na wakacje" is equivalent to the Polish "Jutro jedzie na wakacje" and would be translated as "He is going on vacation tomorrow" [8].

In the Polish language, the plural formation is not made by adding the letter "s" as a suffix to a word, but rather each word has its own plural variant (e.g., "pies-psy", "artysta-artyści", etc.). Additionally, prefixes before nouns like "a", "an", "the", do not exist in Polish (e.g., "a cat-kot", "an apple-jabłko", etc.) [8].

The Polish language has only three tenses (present, past and future). However, it must be noted that the only indication whether an action has ended is an aspect. For example, "Robiłem pranie." Would be translated as "I have been doing laundry", but "Zrobiłem pranie" as "I have done laundry", or "płakać-wypłakać" as "cry-cry out" [8].

The gender of a noun in English does not have any effect on the form of a verb, but it does in Polish. For example, "Zrobił to. – He has done it", "Zrobiła to. – She has done it", "lekarz/lekarka-doctor", "uczeń/uczennica = student", etc. [8].

As a result of this complexity, progress in the development of SMT systems for West-Slavic languages has been substantially slower than for other languages. On the other hand, excellent translation systems have been developed for many popular languages.

5.1.5.3 Spoken vs written language

The differences between speech and text within the context of the literature should also be clarified. Chong [9] pointed out that writing and speech differ considerably in both function and style. Writing tends towards greater precision and detail, whilst speech is often punctuated with repetition and includes prosody, which writing does not possess, to further convey intent and tone beyond the meaning of the words themselves.

According to William Bright [10], spoken language consists of two basic units: Phonemes, units of sound, (that are themselves meaningless) are combined into morphemes, which are meaningful (e.g., the phonemes /b/, /i/, and /t/ form the word "bit"). Contrary alphabetic scripts work in similar way. In a different type of script, the basic unit corresponds to a spoken syllable. In logographic script (e.g., Chinese), each character corresponds to an entire morpheme, which is usually a word [10].

It is possible to convey the same messages in either speech or writing, but spoken language typically conveys more explicit information than writing. The spoken and written forms of a given language tend to correspond to one or more levels and may influence each other (e.g., "through" is spoken as "thru").

In addition, writing can be perceived as colder, or more impersonal, than speech. Spoken languages have dialects varying across geographical areas and social groups. Communication may be formal or casual. In literate societies, writing may be associated with a formal style and speech with a more casual style. Using speech requires simplification, as the average adult can read around 300 words per minute, but the same person would be able to follow only 150–200 spoken words in the same amount of time [11]. That is why speech is usually clearer and more constrained.

The punctuation and layout of written text do not have any spoken equivalent. But it must be noted that some forms of written language (e.g., instant messages or emails) are closer to spoken language. On the other hand, spoken language tends to be rich in repetition, incomplete sentences, corrections and interruptions [12].

When using written texts, it is not possible to receive immediate feedback from the readers. Therefore, it is not possible to rely on context to clarify things. There is more need to explain things clearly and unambiguously than in speech, which is usually a dynamic interaction between two or more people. Context, situation and shared knowledge play a major role in their communication. It allows us to leave information either unsaid or indirectly implied [12].

5.1.5.4 *Machine translation*

Moses is a tool environment for statistical machine translation that enables users to train translation models for any two languages. This implementation of the statistical approach to machine translation is currently the dominant approach in this field of research [13].

The baseline system testing was done using the Moses open source SMT toolkit with its Experiment Management System (EMS) [13]. The SRI Language Modeling Toolkit (SRILM) [14] with an interpolated version of the Kneser-Ney discounting (interpolate –unk –kndiscount) was used for 5-gram language model training. The MGIZA++ tool was used for word and phrase alignment. KenLM [15] was used to binarize (transform features of a text entity into vectors of numbers) the language model, with the lexical reordering set to the msd-bidirectional-fe model [16]. The symmetrization method was set to grow-diag-final-and for word alignment processing [13].

We raised our score in PL-* experiments through changing the language model order from 5 to 6, and also changed the discounting method from Kneser-Ney to Witten-Bell. In the training part, we changed the lexicalized reordering method from msd-bidirectional-fe to hier-mslr-bidirectional-fe. The system was also enriched with Operation Sequence Model (OSM) [17]. The motivation for OSM is that it provides phrase-based SMT models the ability to memorize dependencies and lexical triggers, it can search for any possible reordering, and it has a robust search mechanism. In addition to this, we used Compound Splitting feature [13]. Tuning was done using MERT tool with batch-mira feature and n-best list size was changed from 100 to 150.

Because of a much bigger dictionary, the translation from EN to PL is significantly more complicated. We determined that not using lower casing, changing maximum sentence length to 85, and maximum phrase length to 7 improves the BLEU score. Additionally, we set the language model order from 5 to 6 and changed the discounting method from Kneser-Ney to Witten-Bell. In the training part, we changed the lexicalized reordering method from msd-bidirectional-fe to tgttosrc. The system was also enriched with Operation

Sequence Model (OSM). We also used Compound Splitting feature and did punctuation normalization. Tuning was done using MERT tool with batch-mira feature and n-best list size was changed from 100 to 150. Training a hierarchical phrase-based translation model also improved results in this translation scenario [18].

In PL-CS experiments it was necessary to adjust parameters so that they suit translation between two morphologically rich languages. We changed language model order to 8 which, in contrast with PL-EN translation, produced positive results. The maximum sentence length was changed to 100 from 80 (in PL-EN exceeding 85 did not raise the scores). The lexicalized reordering method was changed to wbe-msd-bidirectional-fe, in order to use word-based data extraction. In tuning, the n-best list size was raised to 200 because there were many possible candidates to be chosen in that language pair.

5.1.5.5 Evaluation

Metrics are necessary in order to measure the quality of translations produced by the SMT systems. For this, various automated metrics are available for comparing SMT translations to high quality human translations. Since each human translator produces a translation with different word choices and orders, the best metrics measure SMT output against multiple reference human translations. For scoring purposes we used four well-known metrics that show high correlation with human judgments. Among the commonly used SMT metrics are: Bilingual Evaluation Understudy (BLEU), the U.S. National Institute of Standards & Technology (NIST) metric, the Metric for Evaluation of Translation with Explicit Ordering (METEOR) and Translation Error Rate (TER). According to Koehn, BLEU [16] uses textual phrases of varying length to match SMT and reference translations. Scoring of this metric is determined by the weighted averages of those matches [19]. To encourage infrequently used word translation, the NIST [19] metric scores the translation of such words higher and uses the arithmetic mean of the n-gram matches. Smaller differences in phrase length incur a smaller brevity penalty. This metric has shown advantages over the BLEU metric. The METEOR [19] metric also changes the brevity penalty used by BLEU, uses the arithmetic mean like NIST, and considers matches in word order through examination of higher order n-grams. These changes increase score based on recall. It also considers best matches against multiple reference translations when evaluating the SMT output. TER [20] compares the SMT and reference translations to determine the minimum number of edits a human would need to make for the translations to be equivalent in both fluency and semantics. The closest match to a reference translation is used in this metric. There are several types of edits considered: Word deletion, word insertion, word order, word substitution and phrase order.

5.1.5.6 Results

The experiment results were gathered in Tables 45, 46 and 47. BASE stands for baseline systems settings and BEST for translation systems with modified training settings (in accordance to Section 4.2). Table 45 contains translation results for written texts and Table 46 for spoken language. The decision was made to use EU Bookshop[31] (EUB) document-based corpus as an example of written language and the QCRI Educational Domain Corpus[32] (QED) (open multilingual collection of subtitles for educational videos and lectures). Both corpora are comparable examples of spoken and written language. The corpora specification is showed in Table 44.

Table 44. Corpora statistics.

Corpus	Language pair	Number of sentences	Number of unique tokens		Avg. sentence length
			PL	FOREIGN	
EUB	PL-EN	537,941	346,417	207,563	23
	PL-CS	435,376	300,897	299,402	22
QED	PL-EN	331,028	168,961	60,839	12
	PL-CS	475,621	172,891	159,325	14

The results presented in Table 44 confirm statements from the Section 3. There is a big dictionary gap between West-Slavic languages and English that is not present within the West-Slavic group itself. We can also observe significant difference in sentence lengths between written and spoken language.

Table 45. Translation of written language.

SYSTEM	Direction	BLEU	NIST	TER	METEOR
BASE	PL->CS	28.93	5.90	67.75	44.39
BEST	PL->CS	29.76	5.98	65.91	45.84
BASE	PL->EN	31.63	6.27	66.93	53.24
BEST	PL->EN	32.81	6.47	65.12	53.92
BASE	CS->PL	26.32	5.11	73.12	42.91
BEST	CS->PL	27.41	5.66	70.63	43.19
BASE	EN->PL	22.18	5.07	74.93	39.13
BEST	EN->PL	23.43	5.33	72.18	41.24

---

[31] http://bookshop.europa.eu
[32] http://alt.qcri.org/resources/qedcorpus/

Table 46. Translation of spoken language.

SYSTEM	Direction	BLEU	NIST	TER	METEOR
BASE	PL->CS	7.11	3.27	79.96	29.23
BEST	PL->CS	8.76	3.95	74.12	32.34
BASE	PL->EN	15.28	5.08	68.45	48.49
BEST	PL->EN	15.64	5.11	68.24	48.75
BASE	CS->PL	6.77	3.12	76.35	29.89
BEST	CS->PL	7.43	3.51	75.76	31,73
BASE	EN->PL	7.76	3.55	79.32	31.04
BEST	EN->PL	8.22	3.78	77.24	32.33

As presented in Tables 45 and 46, it can be observed that translation between West-Slavic and English language usually gives better results that translation within the West-Slavic group. The most likely reason for this is the morphological richness in this group which produces far too many word and phrase mappings, making it statistically difficult choose correct ones. In addition, translation of spoken language produces much poorer results that those of written texts. Some of this difference is obviously the gap between the size of training data but high level of formalisms in written language structures seems to produce the biggest impact. What is more, those formal language structures make it possible to overcome disparities between languages (the difference in scores between PL-CS and PL-EN translation is small, such phenomena is not replicated in spoken language). Because it was observed that in the case of speech there is big difference in translation quality between PL-EN and PL-CS it was decided that additional experimentation on translation from PL to CS would be conducted using EN as a pivot language. Such experiment is showed in Table 47. This idea produced positive results in spoken language translation but negative in written. The most likely reason for this is the fact that in written language translation quality was already similar for both pairs.

Table 47. Translation with EN as pivot language.

CORPUS	Direction	BLEU	NIST	TER	METEOR
EUB	PL->CS	28.93	5.90	67.75	44.39
EUB	PL->EN->CS	26.12	4.77	72.03	41.95
QED	PL->CS	7.11	3.27	79.96	29.23
QED	PL->EN->CS	9.23	4.18	76.33	41.42

5.1.6 SMT in augumented reality systems

An augmented reality (AR) system combines actual objects with computer-generated information in order to enrich the interaction of a user with the real world [224]. When many people think of AR, they think of the historical head-mounted display (HMD) worn by users. However, AR is not limited to display technologies such as the HMD. For example, video from a smart phone can be enhanced with information that is annotated on the video image. The ubiquity and affordability of the smart phone has led many researchers to investigate different AR applications that leverage smart phone capabilities.

We find written text all around us, but in many different languages [153]. A user of an AR system may not know the language of such text. One such application is translating foreign language text in images taken by a smart phone. There are many potential uses for such a capability. For example, it would be useful for a traveler to be able to read signs, posters and restaurant menus when visiting a foreign country. A traveler may also have need of medical attention in another country. Having the ability to read handheld medical documents, translated to their own language, using a smart phone would be a tremendous help to these individuals. In addition, a similar capability for a physician attending to a traveler to read medical documents written in the traveler's language would be very useful [65, 225].

There are several aspects to translating text from a smart phone video image. Image processing techniques must be used to identify text and extract it from the image. This may require various enhancements of the image. Next, optical character recognition (OCR) must be applied to the text in an image to distinguish the characters in a particular language, producing a string of text in that language. Lastly, the text can then be automatically translated from the source language to another target language.

Machine translation (MT), translation by a computer in an automated fashion, is a very challenging step for an AR system. Modern MT systems are not based on the rules that specific languages follow. Statistical machine translation (SMT) systems apply statistical analysis to a very large volume of training data for a pair of languages. These statistics, used to tune an SMT system, can then be used to translate new texts from one of the languages to another. If an SMT system is trained for a particular domain of texts (e.g., road signs or medical texts), then it can produce better results on new texts from that domain [226].

The SMT aspects of an AR system are critical. If the end objective is to augment a display with translated text, then translation is obviously important. A mistranslated road sign could send a traveler in the opposite direction, even resulting in a safety problem in a foreign country. Misunderstandings between a physician and a patient speaking different languages about medical documents could have more serious consequences, such as a misdiagnosis that results in incorrect treatment and corresponding health consequences.

Translation of medical leaflets for patients is another application. Students may find it useful to obtain translated lectures on subjects such as biology and computer science.

The quality of machine translation has greatly improved over time. Several general-purpose translation services, such as Google Translate [227], are currently available on the web. However, they are intended to translate text from a wide variety of domains, and are therefore neither perfect nor well-tuned to produce high quality translations in specific domains.

While our research is extensible to other language pairs, we focus on the challenge of Polish and English. Polish is classified in the Lechitic language group, a branch of the West Slavic language family that includes Czech, Polish and Slovakian. The Polish language uses both animate and inanimate nouns, seven cases and three genders. The agreement of adjectives with nouns is required in number, case and gender. Polish has complex rules for applying these elements, and many words are borrowed from other languages. Its vocabulary is enormous. All this together makes the language a very significant challenge for SMT. Likewise, English is known to present considerable translation challenges, and the syntaxes of the two languages are considerably different. The typical gross mismatch in vocabulary sizes between Polish and English also complicates the SMT process [24].

This monograph will first review the state of the art in machine translation for AR systems. Next, we will describe our research approach, preparation of the language data, and methods of SMT evaluation. Experiments will be described and their results presented. Lastly, we will discuss the results and draw conclusions.

5.1.6.1 *Review of state of the art*

An AR system that enables a user wearing a head-mounted display to interact with a handheld document is presented in [228]. Using their system, a user can select words or very short phrases in the document to access a dictionary. This mechanism could be used to select words or short phrases for translation. However, the focus of their work is on merging a document tracking method with a finger pointing method in order to enable interaction with the handheld document (via the imager worn by the user). Language translation is not the focus of that research, nor is it demonstrated along with their system. In addition, this approach could only translate single words or short phrases that are pre-segmented by the user. The user must select the text for processing.

The authors of [153] describe a system that extracts and translates a word, once designated by a user in an area of an image. Much of its processing runs on a smart phone's processor. However, the focus of their research is image processing. The system requires an Internet connection and uses Google Translate to perform the text translation. "TextGrabber + Translator" [229] is an Android phone-based application that extracts and translates from a variety of printed sources by using a phone's camera. This application also uses Google Translate. Researchers in [230] describe a Microsoft "Snap and Translate" mobile application that extracts text (designated by a user), performs OCR and translates between English and Chinese. This prototype system uses a client and cloud architecture, with Bing Translator as the translation service. The authors do not present translation accuracy results.

The Word Lens translation application for Apple iPhone images is described in [231] and [232]. Accuracy is mentioned as a challenge. Word Lens only translates words and short phrases, not entire texts. EnWo, discussed in [225], is a similar Android application that translates among several languages. Both applications are commercial applications, not open source, which limits their availability. A system that translates text captured by a mobile phone's camera is briefly described in [233]. This system displays overlays of translated words on the phone's camera. However, neither the translation approach nor its accuracy was reported. The focus of this effort appears to be techniques for OCR and augmenting displays for the user.

The authors of [234] describe a prototype Android phone-based system that detects and translates text in the Bangla (a.k.a. Bengali) language. This system uses the Tesseract engine for performing OCR and Google Translate for translation into English. The authors of [224] describe a translator application, iTranslatAR, that translates text in images or manually-entered text on the iOS platform to augment the missing hearing sense and for other applications. This prototype application uses Tesseract OCR and Google Translate as well.

The authors of [235] describe a system that uses genetic algorithms to recognize text in smart phone camera images, such as road signs, and makes it available for translation from English to Tamil through online translation services. However, this research is focused on image processing and reducing the bandwidth required for images. The quality of the translation itself is limited to online translation services. It is also unclear whether or not text at different distances can be processed by this prototype system. In addition, the accuracy of the resulting translations was not evaluated.

A head-mounted text translation system is described in [236]. This research focuses on eye gaze gestures in a head-mounted display and an innovative OCR method. These researchers implemented an embedded Japanese-to-English translation function that is based on common Japanese words translated using Microsoft's Bing translator. However, the accuracy of the translation was not assessed.

A text recognition and translation system hosted on a smart phone and web-based server is discussed in [65]. The user of this system specifies the region of an image, obtained by the phone's camera, that contains text. The image is processed in order to limit the required bandwidth and then transmitted to the server, where text recognition and translation is performed. The prototype system, which focused on English and simplified Chinese as two very challenging languages, uses the open-source Tesseract OCR engine and Google Translate for language translation. Due to the challenge of recognizing text, a user can correct several aspects of character classification in order to improve accuracy. Evaluation showed poor phrase match accuracy (under 50% in all cases).

We observe from this literature that machine translation performance has not been the focus of research for augmented reality systems. Concluding from described examples, most research efforts have emphasized image processing techniques, OCR, mobile-network architecture, or user interaction approaches. This is further indicated by the frequent lack of reporting on translation accuracy results from these studies.

In addition, previous studies have typically used online translation services, scaled-down versions of them, or proprietary software in the language translation portion of their systems. Some of the translation systems used are commercial and proprietary, not open source and not readily available. Even the online translation systems are often not free for AR applications. For example, Google charges based on usage for their Google Translate API as part of the Google Cloud Platform [227]. An online translation service, such as Google, also requires a constant Internet connection for an AR system to perform translation that introduces lags.

The lack of focus on domain-adapted machine translation systems for augmented reality is particularly noteworthy. Such general purpose translation systems collect bilingual data from wherever it is available in order to gather as much data as possible. However, the domain (e.g., medical texts or road signs) to which the resulting translation system is applied may be drastically different from that general data. Our approach is to use in-domain adaptation so that a translation system can be tuned for a specific narrow domain. Research has shown that such adaptation can improve translation accuracy in a specific domain [226, 108]. In-domain adaptation promises to provide similar benefits to AR applications because it also reduces computation costs and is some cases makes it possible to run SMT engine on the handheld device.

5.1.6.2 *The approach*

This research effort developed a tool that recognizes text in a camera image, translates it between two languages, and then augments the camera image (real-time on device screen) with the translated text. As discussed in the review of the state-of-the-art, such tools are already on the market, but most of them translate only single words, like city signs, or work in a less than fully-automated fashion. This work is innovative because it works in real time and translates an entire text, paragraphs or pages. In this manner, the translation's context is maintained, resulting in an improved translation quality. For example, someone might go to a restaurant and view a menu that would be automatically translated on their mobile device.

Translation systems work best if their domain is limited. General purpose translation systems, on the other hand, produce lower-quality translations in very specific domains. For example, a general purpose system such as Google Translation would not do well in translating medical leaflets, since medicine is a very specific and specialized domain.

The approach adopted in this research uses a unique SMT systems trained for specific domains. This should produce higher-quality translations then a general purpose translation system. The domains selected for SMT adaptation are: Medical leaflets, restaurant menus, news, and article or lecture translation.

The tool developed during this research translates all the text it recognizes in the camera image. Most tools from the reviewed literature translate single words or small, selected phrases. This results in those approaches losing the context of the translation. Loss of that context can only lead to lower quality. In addition, our approach is easy to implement and also based on open source tools, not restricted in use, as are commercial solutions. It is also extensible to any language pair, not just limited to the languages used in our experiments.

5.1.6.3 Data preparation

This research used Polish and English language data from two primary and distinct parallel corpora: The European Medicines Agency (EMEA) and the 2013 Technology, Entertainment, Design (TED) lectures. The EMEA data is composed of approximately 1,500 biomedical documents in Portable Document Format (PDF)—approximately 80MB of data in UTF-8 format—that relate to medicinal products. These documents are translated into the 22 official European Union languages. Separation of these texts into sentences yielded 1,044,764 sentences, which were constructed from 11.67 M untokenized words. The disproportionate Polish-English vocabulary in the EMEA data is composed of 148,170 and 109,326 unique Polish and English words, respectively.

Errors in the Polish EMEA data first had to be repaired. Additional processing of the Polish data, which used the Moses SMT processing toolkit [8], was required prior to its use in training the SMT model. This toolkit includes tools for creating, training, tuning and testing SMT models. Removal of long sentences (defined as 80 tokens) was part of this preprocessing. Preparation of the English data was much simpler than that for Polish, and there were fewer errors in the data. However, the removal of strange UTF-8 symbols and foreign words was required.

The TED lecture data, totaling approximately 17 MB, included approximately 2.5 M untokenized Polish words, which were separated into sentences and aligned in language pairs. Similar to [18], the data was preprocessed using both automated cleaning and manual changes in order to correct problems, including spelling errors, nesting problems (described in [18]) and repeated text fragments that caused issues in text parallelism. The authors describe the tool used to correct many of these problems in [24]. Preparation of the English data was, once again, much simpler than that of the Polish data. The main focus was cleaning of the data to eliminate errors.

As in the EMEA data, this processing of TED lecture data resulted in a disproportionate vocabulary of 59,296 and 144,115 unique English and Polish words, respectively. However, the domain of the TED lectures was much less specific and more wide-ranging then that of the EMEA data.

5.1.6.4 *Experiments*

Experiments were conducted to evaluate different SMT systems using various data. In general, each corpus was tokenized, cleaned, factorized, converted to lowercase, and split. A final cleaning was performed. For each experiment, the SMT system was trained, a language model was applied and tuning was performed. The experiments were then performed. For OCR purposes we used well known from good quality Tesseract engine [140], by this we also evaluated an impact of OCR mistakes on translation. It also must be noted that the OCR system was not adapted to any specific types of images or texts which would most probably improve its quality. It was not done because it was not goal of this research.

The Moses toolkit, its Experiment Management System and the KenLM language modeling library [14] were used to train the SMT systems and conduct the experiments. Training included use of a 5-gram language model based on Kneser-Ney discounting. SyMGIZA++ [237], a multi-threaded and symmetrized version of the popular GIZA++ tool [8], was used to apply a symmetrizing method in order to ensure appropriate word alignment. Two-way alignments were obtained and structured, leaving the alignment points that appear in both alignments. Additional points of alignment that appear in their union were then combined. The points of alignment between the words in which at least one was unaligned were then combined (grow-diag-final). This approach facilitates an optimal determination of points of alignment [11]. The language model was binarized by applying the KenLM tool [234].

Finally, the approach described by Durrani et al. in [80] was applied to address out-of-vocabulary (OOV) words found in our test data. Despite the use of large amounts of language data, OOV words, such as technical terms and named entities, pose a challenge for SMT systems. The method adopted here from [80] is a completely unsupervised and language-independent approach to OOV words.

Using traditional SMT systems, experiments were performed on EMEA medical leaflet data and then on TED lecture data, translating Polish to English. These experiments were repeated for texts taken randomly from the corpora as well as on the same texts recognized by the OCR system. Data was obtained through the process of random selection of 1,000 sentences and their removal from the corpora. The BLEU, NIST, TER and METEOR metrics were used to evaluate the experimental results.

Tables 48 and 49 provide the experimental results for translation. The EMEA abbreviation stands for translation of medical leaflets, TED for discussed earlier TED corpus and FIELD for field experiments that were performed on random samples of Polish city signs, posters, restaurant menus, lectures on biology and computer science, and medicine boxes. The field SMT system was trained on both TED and EMEA corpora. The images were taken in different situations, lightning and devices. Experiments that have additional "–OCR" suffix in their name were first processed by OCR engine and secondly translated.

Table 48. Polish-to-English translation.

System	BLEU	NIST	METEOR	TER
EMEA	76.34	10.99	85.17	24.77
EMEA-OCR	73.58	10.77	79.04	27.62
TED	28.62	6.71	58.48	57.10
TED-OCR	27.41	6.51	56.23	60.02
FIELD	37,41	7,89	63,76	45,31
FIELD-OCR	35,39	7,63	61,23	48,14

Table 49. English-to-Polish translation.

System	BLEU	NIST	METEOR	TER
EMEA	73.32	10.48	81.72	27.05
EMEA-OCR	69.15	9.89	75.91	29.45
TED	26.61	5.99	48.44	59.94
TED-OCR	23.74	5.73	46.18	61.23
FIELD	33,76	7,14	61,68	49,56
FIELD-OCR	32,11	6,99	59,17	52,05

Translation phrase tables often grow enormously large, because they contain a lot of noisy data and store many very unlikely hypotheses. This is problematic when a real-time translation system that requires loading into memory is required, especially on mobile devices with limited resources. We decided to try a method of pruning those tables introduced by Johnson et al. [126] and our method based on dictionary.

Table 50. Pruning results.

Optimizer	BLEU	Phrase Table (GB)	Reordering Table (GB)
None	**35.39**	**6.4**	**2.3**
Absolute 3	26.95	1.1	0.4
Absolute 2	30.53	2.5	0.9
Absolute 1	32.07	4.9	1.7
Relative 2.5	30.82	3.1	1.1
Relative 5	26.35	1.1	0.4
Relative 7.5	17.68	0.3	0.1
Pruning 30	**32.36**	**1.9**	**0.7**
Pruning 60	32.12	2.0	0.7
Pruning 90	32.11	2.0	0.75
Pruning 20	32.44	2.1	0.75
Absolute 1 + Pruning 20	**30.29**	**0.85**	**0.3**

In Table 50, experiments are shown (Absolute N—keep sentence if at least N words from dictionary appear in it, relative N—keep a sentence if at least N% of sentence is built by from dictionary, pruning N—keep only N most probable translations). We conducted those experiments on the most important translation engine which is the field test starting with OCR from Polish to English language.

Contrary to the referenced publications, the empirical experiments showed that pruning (for PL-EN) decreases translation quality significantly. Even though phrase and reordering tables are much smaller, the decrease in quality is questionable. What is interesting is that a substantial factor of quality loss to final phrase table size was obtained by combining pre-filtering and pruning of a phrase table in Experiment 12.

5.1.6.5 Discussion

AR systems would greatly benefit from the application of state-of-the-art SMT systems. For example, translation and display of text in a smart phone image would enable a traveler to read medical documents, signs and restaurant menus in a foreign language. In this monograph, we have reviewed the state of the art in AR systems, described our SMT research approach—including the important preparation of language data and models, as well as evaluation methods—and presented experimental results for several different variants of Polish-English SMT systems.

Clearly, machine translation performance has not been the focus of AR research. The literature shows an emphasis on image processing techniques, OCR, mobile-network architecture and user interaction approaches. As a result, previous AR research has generally used general-purpose translation services, which are not tuned for specific text domains and do not produce the highest quality translations. In addition, AR research reported in the literature frequently lacks reporting of translation accuracy results.

The focus of this research was machine translation between the Polish and English languages. Both languages present significant challenges due to vast differences in syntax, language rules, and vocabulary. Experiments were conducted in Polish-to-English and English-to-Polish direction on the EMEA and TED data. In addition, field experiments were performed on a variety of random samples in Polish. Standard machine translation evaluation methods were used to evaluate the results, which show great promise for the eventual SMT use in AR systems.

BLEU scores lower than 15 mean that the machine translation engine is unable to provide satisfactory quality, as reported by Lavie [144] and a commercial software manufacturer [119]. A high level of post-editing will be required in order to finalize output translations and reach publishable quality. A system score greater than 30% means that translations should be understandable without problems. Scores over 50 reflect good and fluent translations.

Overall, the EMEA and TED experimental results appear adequate within the limits of the specific text domain. SMT is particularly useful for the complex Polish language. However, there is clearly room for future improvement for critical applications. English-to-Polish translation proved more challenging, as expected. The results from the OCR experiment generally show translation performance only slightly lower then plain SMT. The most likely reason for this is that statistical translation methods are not vulnerable to small mistakes like spelling, etc. All the results were very encouraging, but more work remains to be done in order to further optimize SMT translation quality. The results also reinforce the importance of phrase table pruning so it can be small enough to fit into mobile device memory. In our experiments, we were able to obtain such threshold by reducing quality by about 5 BLEU points.

There are many advantages to the SMT approach that we took in our research. First, an advanced SMT system was adapted for particular text domains. Such a domain-adapted system produces higher-quality translations then general-purpose translation services, such as Google Translate, in specific domains. In addition, our approach is not restricted by commercial license and is easy to implement.

The proposed SMT approach, unlike other AR systems, translates all the text it sees without need to transfer data over Internet. This promotes automation and flexibility in an AR system user, since they do not have to identify specific text for translation and are not limited to simple words or phrases. Translating larger texts also leverages the value of context, which increases translation quality. Lastly, our approach is easily extensible to any language pair.

5.1.7 Neural networks in machine translation

The quality of machine translation is rapidly evolving. Today one can find several machine translation systems on the web that provide reasonable translations, although the systems are not perfect. In some specific domains, the quality may decrease. A recently proposed approach to this domain is neural machine translation. It aims at building a jointly-tuned single neural network that maximizes translation performance, a very different approach from traditional statistical machine translation. Recently proposed neural machine translation models often belong to the encoder-decoder family, in which a source sentence is encoded into a fixed length vector that is, in turn, decoded in order to generate a translation. This part of the research examines the effects of different training methods on a Polish-English Machine Translation system used for medical data. The European Medicines Agency parallel text corpus was used as the basis for training of neural and statistical network-based translation systems. A comparison and implementation of a medical translator is the main focus of this experiments.

MT systems have no knowledge of language rules. Instead, they translate by analyzing large amounts of text in language pairs. They can be trained for specific domains or applications using additional data germane to a selected domain. MT systems typically deliver translations that sound fluent, although they tend to be less consistent than human translations. Statistical machine translation (SMT) has rapidly evolved in recent years. However, existing SMT systems are far from perfect, and their quality decreases significantly in specific domains. For instance, Google Translate would work well for generalized texts like dialogs or traveler aid but would fail when translating medical or law terms. To show the contrast, we will compare the results to Google Translate engine. The scientific community has been engaged in SMT research. Among the greatest advantages of statistical machine translation is that perfect translations are not required for most applications [8]. Users are generally interested in obtaining a rough idea of a text's topic or what it means.

However, some applications require much more than this. For example, the beauty and correctness of writing may not be important in the medical field, but the adequacy and precision of the translated message is very important. A communication or translation error between a patient and a physician in regard to a diagnosis may have serious consequences on the patient's health. Progress in SMT research has recently slowed down. As a result, new translation methods are needed. Neural networks provide a promising approach for translation [139] due to the their rapid progress in terms of methodology and computation power. They also bring the opportunity to overcome limits of statistical-based methods that are not context-aware.

Machine translation has been applied to the medical domain due to the recent growth in the interest in and success of language technologies. As an example, a study was done on local and national public health websites in the USA with an analysis of the feasibility of edited machine translations for health promotional documents [173]. It was previously assumed that machine translation was not able to deliver high quality documents that can used for official purposes. However, language technologies have been steadily advancing in quality. In the not-too-distant future, it is expected that machine translation will be capable of translating any text in any domain at the required quality.

The medical data field is a bit narrow, but very relevant and a promising research area for language technologies. Medical records can be translated by use of machine translation systems. Access to translations of a foreign patient's medical data might even save their life. Direct speech-to-speech translation systems are also possible. An automated speech recognition (ASR) system can be used to recognize a foreign patient's speech. After it is recognized, the speech could be translated into another language with synmonograph in real time. As an example, the EU-BRIDGE project intends to develop automatic transcription and translation technology. The project desires innovative multimedia translation services for audiovisual materials between European and non-European languages.[33]

Making medical information understandable is relevant to both physicians and patients [238]. As an example, Healthcare Technologies for the World Traveler emphasizes that a foreign patient may need a description and explanation of their diagnosis, along with a related and comprehensive set of information. In most countries, residents and immigrants communicate in languages other than the official one [239].

Karliner et al. [67] talks of the necessity of human translators obtaining access to healthcare information and, in turn, improving its quality. However,

---

[33] http://www.eu-bridge.eu

telemedicine information translators are often not available for either professionals or patients. Machine translation should be further developed in order to reduce the cost associated with medical translation [68]. In addition, it is important to increase its availability and overall quality.

Patients, medical professionals and researchers need adequate access to the telemedicine information that is abundant on the web [70]. This information has the potential to improve health and well-being. Medical research could also improve the sharing of medical information. English is the most used language in medical science, though not the only one.

The goal of this research is to present the experiments on neural-based machine translation in comparison to statistical machine translation. The adaptation of translation techniques, as well as proper data preparation for the need of PL-EN translation, was also necessary. Such systems could possibly be used in real time speech to speech translation systems and aid foreign travelers that would require medical assistance. Combining such system with OCR and an augmented reality tool could bring a real time translator to mobile devices as well. Human interpreters with a proper medical training are extremely rare and costly. Machine translation could also assist in the evaluation of medical history, diagnoses, proper medical treatments, health related information and the findings of medical researchers from the entire word. Mobile devices, the Internet and web applications can be used to boost delivery of machine translation services for medical purposes, even as real time speech-to-speech services.

5.1.7.1 Neural networks in translations

Machine learning is the programming of computers for optimization of a performance criterion with the use of past experience/example data. It focuses on the learning part of conventional intelligence. Based on the input type available while training or the desired outcome of the algorithms, machine learning algorithms can be organized into different categories, e.g., reinforced learning, supervised learning, semi-supervised learning, unsupervised learning, development learning and transductive inference. With a number of algorithms, learning of each type can take place. The artificial neural network (ANN) is a unique learning algorithm inspired by the functional aspects and structure of the brain's biological neural networks. With use of ANN, it is possible to execute a number of tasks, such as classification, clustering and prediction, using machine learning techniques like supervised or reinforced learning. Therefore, ANN is a subset of machine learning algorithms [192].

Neural machine translation is a new approach to machine translation in which a large neural network is trained to maximize translation performance. This is, undoubtedly, a radical departure from existing phrase-based statistical translation approaches, in which a translation system consists of subcomponents that are separately optimized. A bidirectional recurrent neural network (RNN), known as an encoder, is used by the neural network in order to encode a source sentence for a second RNN, known as a decoder, that is used to predict words in the target language [240].

Now, let us compare neural networks with human activity. Whenever a new neural network is created, it is like a child being born. After birth, the training of the network starts. Unsurprisingly, it is possible to create the neural networks for certain domains or applications. Here, the difference between neural networks and childbirth is rather obvious. First, a decision to create the needed neural network is made. On the other hand, childbirth results are naturally random. After childbirth, one cannot know whether a child will concentrate on studies through his or her entire life. That is left in the hands of the child and the parents. Parents undoubtedly play an important role in child development, and this aspect is similar to the person creating a neural network. In the same way that a child develops to be an expert in a certain domain, neural networks are also trained to be experts in specific domains. Once an automatic learning mechanism is established in neural networks, it practices, and with time it does work on its own, as expected. After proof that the neural network is working correctly, it is called an "expert," as it operates according to its own judgment [241].

Both the human and neural networks can learn and develop to be experts in a certain domain, with both being mortal. But what is the difference between them? The major difference is that humans can forget, unlike neural networks. A neural network will never forget, once trained, as information learned is permanent and hard-coded. Human knowledge may not be permanent. A number of factors may cause the death of brain cells, with stored information getting lost, and the human brain starts forgetting [241].

Another major difference is accuracy. Once a particular process is automated via a neural network, the results can be repeated and will definitely remain as accurate as calculated the first time. Humans are rather different. The first ten processes may be accurate, with the next ten characterized by mistakes. Another major difference is speed [241].

5.1.7.2 Data preparation

The used corpus was derived from the European Medicines Agency (EMEA[34]) parallel corpus with Polish language data included. It was created from EMEA's biomedical PDF documents. The derived corpus includes medical

---

[34] http://opus.lingfil.uu.se/EMEA.php

product documents and their translation into 22 official European Union languages [242]. It consists of roughly 1,500 documents for each language, but not all of them are available in every language. The data comprises 80 MB and 1,044,764 sentences constructed from 11.67 million untokenized words.[35] The data is UTF-8 encoded text. In addition, the texts were separated and structured in language pairs. This corpus was chosen as the most similar to medical texts (which could not be accessed in sufficient quantity) in terms of vocabulary and complexity.

The vocabulary consisted of 109,320 unique English and 148,160 unique Polish tokens. When it comes to translation from English to Polish, the disproportionate vocabulary sizes and the number of tokens makes it a challenging task.

Prior to using the training translation model, preprocessing included long sentence (set to 80 tokens) removal to limit model size and computation time. For this purpose, Moses toolkit scripts were employed [8]. Moses is an open source SMT toolkit supporting linguistically-motivated factors, network decoding and data formats required for efficient use by language and translation models. The toolkit also included an SMT decoder, a wide variety of training tools, tuning tools and a system applied to a number of translation tasks.

English data preparation was much less complicated as compared to that of Polish data. A tool to clean the English data was developed by eliminating strange symbols, foreign words, etc. The English data had much fewer errors, though there was a need to fix some problems. Translations in languages other than English proved to be problematic, with repetitions, UTF-8 symbols and unfinished sentences. When corpora are built by automatic tools, such errors are typical.

5.1.7.3 Translation systems

Various experiments have been carried out for the evaluation of different versions of translation systems. The experiments included a number of steps, including corpora processing, cleaning, tokenization, factorization, splitting, lower casing and final cleaning. A Moses-based SMT system was used for testing with comparison of performance to a neural network.

The Experiment Management System (EMS) [46] was used together with the SMT Moses Open Source toolkit. In addition, the SRI Language Modelling Toolkit (SRILM) [9], used alongside an interpolated version of Kneser-Ney discounting, was employed for training of a 5-gram language model. For

---

[35] https://en.wikipedia.org/wiki/Tokenization_(lexical_analysis)

word and phrase alignment, the MGIZA++ tool [9], a multithreaded version of the GIZA++ tool, was employed. KenLM [10] was employed in order to ensure high quality binaries of the language model. Lexical reordering was placed at the mid-bidirectional-fe model, and the phrase probabilities were reordered according to their lexical values. This included three unique reordering orientation types applied to source and target phrases: Monotone (M), swap (S) and discontinuous (D). The bidirectional model's reordering includes the probabilities of positions in relation to the actual and subsequent phrases. The probability distribution of English phrases is evaluated by "e," and foreign phrase distribution by "f". For appropriate word alignment, a method of symmetrizing the text was developed. At first, two-way alignments from GIZA++ were structured, which resulted in leaving only the points of alignments appearing in both. The next phase involved combination of additional alignment points appearing in the union. Additional steps contributed to potential point alignment of neighboring and unaligned words. Neighboring can be positioned to the left or right, top or bottom, with an additional diagonal (grow dialog) position. In the final phase, a combination of alignment points between words is considered, where some are unaligned. The application of the grow dialog method will determine points of alignment between two unaligned words [18].

The neural network was implemented using the Groundhog and Theano tools. Most of the neural machine translation models being proposed belong to the encoder-decoder family [243] with use of an encoder and a decoder for every language, or use of a language-specific encoder to each sentence application whose outputs are compared [243]. A translation is the output a decoder gives from the encoded vector. The entire encoder-decoder system, consisting of an encoder and decoder for each language pair, is jointly trained for maximization of the correct translation.

A potential downside to this approach is that a neural network will need to have the capability of compressing all necessary information from a source sentence into a vector of a fixed length. A challenge in dealing with long sentences may arise. Cho et al. showed that an increase in length of an input sentence will result in the deterioration of basic encoder performance [243].

That is the reason behind the encoder-decoder model, which learns to jointly translate and align. After generating a word when translating, the model searches for position sets in the source sentence that contains all the required information. A target word is then predicted by the model based on context vectors [78].

A significant and unique feature of this model approach is that it does not attempt to encode an input sentence into a vector of fixed length. Instead, the sentence is mapped to a vector sequence, and the model adaptively chooses a vector subset as it decodes the translation. This gets rid of the burden of a neural translation model compressing all source sentence information, regardless of length, into a fixed length vector [78].

Two types of neural models were trained. The first is an RNN Encoder–Decoder [78], and the other is the model proposed in [78]. Each model was trained with the sentences of length up to 50 words. The encoder and decoder had 1000 hidden units. Multilayer network with a single maxout [244] hidden layer to compute the conditional probability of each target word [245] was used. In addition, a minibatch stochastic gradient descent (SGD) algorithm together with Adadelta [246] was used to train each model. Each SGD update direction was computed using a minibatch of 80 sentences. When models were trained, a beam search algorithm was used in order to find a translation that approximately maximizes the conditional probability [247, 248].

5.1.7.4 *Results*

The experiments were performed in order to evaluate the optimal translation methods for English to Polish and vice versa. The experiments involved the running of a number of tests with use of the developed language data. Random selection was used for data collection, accumulating 1000 sentences for each case. Sentences composed of 50 words or fewer were used, due to hardware limits, with 500,000 training iterations and neural networks having 750 hidden layers. The NIST, BLEU, TER and METEOR metrics were used for evaluation of the results. The TER metric tool is considered as the best one, showing a low value for high quality, while other metrics use high scores to indicate high quality. For comprehension and comparison, all metrics were made to fit into the 0 to 100 range.

Scores lower than 15 BLEU points mean that the machine translation engine is unable to provide satisfactory quality, as reported by Lavie [144] and a commercial software manufacturer.[36] A high level of post-editing will be required in order to finalize output translations and reach publishable quality. A system score greater than 30 means that translations should be understandable without problems. Scores over 50 reflect good and fluent translations.

The computation time for statistical model was 1 day, whereas for each neural model it was about 4–5 days. Statistical model was computed using single Intel Core i7 CPU (8 threads). For neural model calculation the power of GPU units was used (one GeForce 980 card).

The results presented in Table 51 are for Polish-to-English and in Table 52 are for English-to-Polish translation results. Statistical translation

---

[36] http://www.kantanmt.com/

results are annotated as SMT in the tables. Translation results from the most popular neural model are annotated as ENDEC, and SEARCH indicates the neural network-trained systems. The results are visualized on the Figures 33 and 34.

Table 51. Polish-to-English and translation results.

Polish-to-English				
System	BLEU	NIST	METEOR	TER
Google	18,11	36,89	66,14	62,76
SMT	36,73	55,81	60,01	60,94
ENDEC	21,43	35,23	47,10	47,17
SEARCH	24,32	42,15	56,23	51,78

Table 52. English-to-Polish and translation results.

English-to-Polish				
System	BLEU	NIST	METEOR	TER
Google	12,54	24,74	55,73	67,14
SMT	25,74	43,68	58,08	53,42
ENDEC	15,96	31,70	62,10	42,14
SEARCH	17,50	36,03	64,36	48,46

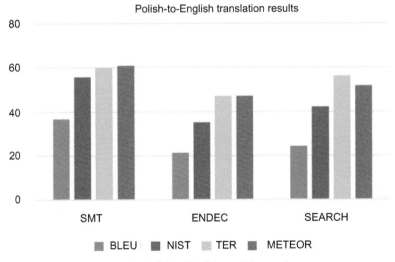

Figure 33. Polish-to-English translation results.

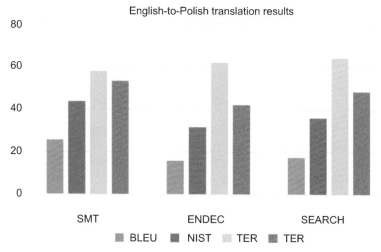

Figure 34. English-to-Polish translation results.

A number of conclusions can be drawn directly from the results of the research experiments. As anticipated, results of the encoder-decoder did not match the SEARCH results. On the other hand, the statistical approach obtained slightly better results than the SEARCH. Nevertheless, it must be noted that a neural network translation system requires fewer resources for training and maintenance. In addition, the translation results were manually analyzed, due to similarity between the neural network and the brain. In many cases, the neural network substituted words with other words occurring in a similar context. As an example, an input sentence "I can't hear you very well" was first translated to "nie za dobrze pana słyszę" then translated to "słabo pana słyszę" meaning "I hear you poorly." Due to the need to preserve meaning, this leads to a conclusion that an automatic statistical evaluation method is not suited to neural machine translation and, also, that the final score should be above the one measured. Neural machine translation shows promise for future automatic translation systems. Even though only a few steps have been taken in this research, satisfactory results are already being obtained. With an enlarged number of hidden layers and an increase in the training iteration numbers, language models have the potential to greatly improve in quality [249]. The availability of GPUs can enhance neural network training so as to become more computationally feasible.

Such systems can also be prepared for any required language pair. Using machine translation on medical texts can have a great potential of ensuring the benefits to patients, including tourists and people who do not know the language of the country in which they require medical help. Improved access to medical information can be very profitable for patients, medical professionals and, eventually, to the medical researchers themselves.

5.1.8 Joint conclusions on SMT

Several conclusions can be drawn from the experimental results presented here. Automatic and manual cleaning of the training files has some positive impact, among the variations of the experiments. Obtaining and adapting additional bilingual and monolingual data produced the biggest influence on the translation quality itself. In each translation direction, using OSM and adapting training and tuning parameters was necessary. This could not be simply replicated from other experiments. It is necessary to adjust translation parameters and techniques separately not only for each language and translation direction but also for each text domain. Hierarchical translation models seem to be beneficial in PL-EN translation.

In addition, converting Polish surface forms of words to lemma reduces the Polish vocabulary, which should improve translation performance. Polish-to-English translation typically outscores English-to-Polish translation, even on the same data. This is also what would be expected in conducted experiments with lemma. Nonetheless, the initial assumptions were not confirmed in empirical tests.

Several potential opportunities for future work are of interest. Additional experiments using extended language models are warranted in order to determine if this improves SMT scores. It is also of interest to develop some more web crawlers so as to obtain additional data that would most likely prove useful. It was observed that the Wikipedia corpus that was created is still very noisy, even after filtering and improvements to Yalign.

In the future, it is intended to try clustering the training data into word classes in order to obtain smoother distributions and better generalizations. Using class-based models was shown to be useful when translating into morphologically-rich languages like Polish [79]. It is planned to use unsupervised transliteration models, which have proved to be quite useful in MT for translating OOV words, for disambiguation, and for translating closely-related languages [80]. Using a fill-up combination technique (instead of interpolation) may be useful when the relevance of the models is known *a priori*—typically, when one is trained on in-domain data and the others on out-of-domain data [57].

Results and Conclusions 159

5.2 Evaluation of obtained comparable corpora

Parallel sentences are a relatively scarce but extremely useful resource for many applications, including cross-lingual retrieval and statistical machine translation. This section evaluates data from newly obtained comparable corpora using the methodologies described in Section 3.2. The evaluation is highly practical, since mined parallel corpora are assessed for performance in machine translation systems for the Polish-English language pair and in various text domains.

5.2.1 Initial mining results using native Yalign solution

To train the classifier, good quality parallel data were needed, as well as a dictionary that included translation probability. For this purpose, the TED talks [15] corpora enhanced during the IWSLT'13 Evaluation Campaign [18] was used, together with BTEC, EMEA, EUP and OPEN corpora. To obtain a dictionary, a phrase table was trained and 1-grams were extracted from it. The MGIZA++ tool for word and phrase alignment was used. The lexical reordering was set to use the msd-bidirectional-fe method [22]. The grow-diag-final-and word alignment symmetrization method was used. This method starts with the intersection of two alignments, growing it in all directions with neighboring and non-neighboring alignment points between words [22, 60, 127]. It is described in more detail in Section 4.9.2. The previously-described data was used for bilingual classifier training. The mining procedure was repeated with each of five different classifiers.

Using this method, it was possible to successfully mine a corpora of approximately 80 MB from Wikipedia and 0.3 MB (1,417 bi-sentences) from Euronews. The parallel data sets were combined into one large corpus for use in the MT experiments. The detailed results for Wikipedia are presented in Table 43. The results in Table 53 are divided in two sets obtained using different classifiers (trained on different corpora). The classifier is SVM described in Section 3.2.1.

During this empirical research, it was realized that Yalign suffers from a problem that produces different results and quality measures, depending on whether the system was trained from a foreign to a native language or vice versa. To cover as much parallel data as possible during the mining, it is necessary to train the classifiers bidirectionally for the language pairs of interest. By doing so, additional bi-sentences can be found. Some of them will be repeated; however, the potential increase in the parallel corpora is worth the effort. Table 54 shows the number of sentences that were obtained in this second phase of the mining, how many of them overlapped, and the number of additional sentences obtained from mining.

Table 53. Data mined from Wikipedia for each classifier.

Classifier	Value	PL	EN
TED	Size in MB	41.0	41.2
	No. of sentences	357,931	357,931
	No. of words	5,677,504	6,372,017
	No. of unique words	812,370	741,463
BTEC	Size in MB	3.2	3.2
	No. of sentences	41,737	41,737
	No. of words	439,550	473,084
	No. of unique words	139,454	127,820
EMEA	Size in MB	0.15	0.14
	No. of sentences	1,507	1,507
	No. of words	18,301	21,616
	No. of unique words	7,162	5,352
EUP	Size in MB	8.0	8.1
	No. of sentences	74,295	74,295
	No. of words	1,118,167	1,203,307
	No. of unique words	257,338	242,899
OPEN	Size in MB	5.8	5.7
	No. of sentences	25,704	25,704
	No. of words	779,420	854,106
	No. of unique words	219,965	198,599

Table 54. Corpora statistics obtained in second mining phase.

Classifier	Value	Data Mined
TED	Recognized sentences	132,611
	Overlapping sentences	61,276
	Newly obtained	71,335
BTEC	Recognized sentences	12,447
	Overlapping sentences	9,334
	Newly obtained	3,113
EMEA	Recognized sentences	762
	Overlapping sentences	683
	Newly obtained	79
EUP	Recognized sentences	23, 952
	Overlapping sentences	21,304
	Newly obtained	2,648
OPEN	Recognized sentences	11,751
	Overlapping sentences	7,936
	Newly obtained	3,815

5.2.2 *Native Yalign method evaluation*

The results of the experiments involving the native Yalign method are shown in Tables 55 and 56. Starting from the baseline system (BASE) tests in the PL to EN and EN to PL directions, evaluation scores were increased by extending the language model (LM), interpolating it (ILM), supplementing the corpora with additional data (EXT), and filtering additional data with Modified Moore-Lewis filtering (MML) [82].

Table 55. Polish-to-English MT experiments.

Corpus	System	BLEU	NIST	TER	METEOR
TED	BASE	16.96	5.26	67.10	49.42
	EXT	16.96	5.29	66.53	49.66
	MML	16.84	5.25	67.55	49.31
	LM	17.14	5.27	67.66	49.95
	ILM	17.64	5.48	64.35	51.19
BTEC	BASE	11.20	3.38	77.35	33.20
	EXT	12.96	3.72	74.58	38.69
	MML	12.80	3.71	76.12	38.40
	LM	13.23	3.78	75.68	39.16
	ILM	13.60	3.88	74.96	39.94
EMEA	BASE	62.60	10.19	36.06	77.48
	EXT	62.41	10.18	36.15	77.27
	MML	62.72	10.24	35.98	77.47
	LM	62.90	10.24	35.73	77.63
	ILM	62.93	10.27	35.48	77.87
EUP	BASE	36.73	8.38	47.10	70.94
	EXT	36.16	8.24	47.89	70.37
	MML	36.66	8.32	47.25	70.65
	LM	36.69	8.34	47.13	70.67
	ILM	36.72	8.34	47.28	70.79
OPEN	BASE	64.54	9.61	32.38	77.29
	EXT	65.49	9.73	32.49	77.27
	MML	65.16	9.62	33.79	76.45
	LM	65.53	9.70	32.94	77.00
	ILM	65.87	9.74	32.89	77.08

To be more specific, the BLEU, METEOR and TER results for the TED corpus, found in Tables 45 and 46, were evaluated in order to determine if the differences in methods were relevant. The variance due to the BASE and MML set selection was measured. It was calculated using bootstrap resampling[37] for each test run. The variances were 0.5 for BLEU, 0.3 for METEOR, and 0.6 for TER. The positive variances mean that there are significant (to some extent) differences between the test sets. It also indicates that a difference of this magnitude is likely to be generated again by some random translation process, which would most likely lead to better translation results in general [83].

Table 56. English-to-Polish MT experiments.

Corpus	System	BLEU	NIST	TER	METEOR
TED	BASE	10.99	3.95	74.87	33.64
	EXT	10.86	3.84	75.67	33.80
	MML	11.01	3.97	74.12	33.77
	LM	11.54	4.01	73.93	34.12
	ILM	11.86	4.14	73.12	34.23
BTEC	BASE	8.66	2.73	85.27	27.22
	EXT	8.46	2.71	84.45	27.14
	MML	8.50	2.74	83.84	27.30
	LM	8.76	2.78	82.30	27.39
	ILM	9.13	2.86	82.65	28.29
EMEA	BASE	56.39	9.41	40.88	70.38
	EXT	55.61	9.28	42.15	69.47
	MML	55.52	9.26	42.18	69.23
	LM	55.38	9.23	42.58	69.10
	ILM	55.62	9.30	42.05	69.61
EUP	BASE	25.74	6.54	58.08	48.46
	EXT	24.93	6.38	59.40	47.44
	MML	24.88	6.38	59.34	47.40
	LM	24.64	6.33	59.74	47.24
	ILM	24.94	6.41	59.27	47.64
OPEN	BASE	31.55	5.46	62.24	47.47
	EXT	31.49	5.46	62.06	47.26
	MML	31.33	5.46	62.13	47.31
	LM	31.22	5.46	62.61	47.29
	ILM	31.39	5.46	62.43	47.33

To verify this analysis, an SMT system using only data extracted from comparable corpora (not using the original domain data) was trained. The

---

[37] https://github.com/jhclark/multeval

mined data was also used as a language model. The evaluation was conducted on the same test sets that were used in Tables 55 and 56. This was done intentionally in order to determine how such a system would cope with the translation of domain-specific text samples. Doing so provided an assessment of the influence of additional data on translation quality and of the similarity between mined data and in-domain data. The results are presented in Tables 57 and 58. The BASE rows show the results for baseline systems trained on original, in-domain data; the MONO rows indicate systems trained only on mined data in one direction; and the BI rows show results for a system trained on data mined in two directions with duplicate segments removed.

The results for the SMT systems based only on mined data are not very surprising. First, they confirm the quality and high level of parallelism of the corpora. This can be concluded from the translation quality, especially for the TED data set. Only two BLEU scoring anomalies were observed when comparing systems strictly trained on in-domain (TED) data and mined data for EN-to-PL translation. It also seems reasonable that the best SMT scores were obtained on TED data. This data set is the most similar to Wikipedia articles, overlapping with it on many topics. In addition, the Yalign classifier trained on the TED data set recognized most of the parallel sentences. The results show that the METEOR metric, in some cases, increases when the other metrics decrease. The most likely explanation for this is that other metrics suffer, in comparison to METEOR, from the lack of a scoring mechanism for synonyms.

Wikipedia is a very wide domain, not only in terms of its topics, but also its vocabulary. This leads to the conclusion that mined corpora are a good source for extending sparse text domains. It is also the reason why test sets originating from wide domains outscore those of narrow domains and, also, why training on a larger mined data set sometimes slightly decreases the results on very specific domains. Nonetheless, in many cases, after manual analysis was conducted, the translations were good but the automatic metrics were lower due to the usage of synonyms. The results also confirm that bi-directional mining has a positive influence on the output corpora.

Bi-sentence extraction has become more and more popular in unsupervised learning for numerous specific tasks. This method overcomes disparities between English and Polish or any other West-Slavic language. It is a language-independent method that can easily be adjusted to a new environment, and it only requires parallel corpora for initial training. The experiments show that the method performs well. The resulting corpora increased MT quality in wide text domains. Decreased or very small score differences in narrow domains are understandable, because a wide text domain such as Wikipedia most likely adds unnecessary n-grams that do not exist in test sets from a very specific domain. Nonetheless, it can be assumed that even small differences can make a positive influence on real-life, rare translation scenarios.

Table 57. PL-to-EN translation results using bi-directional mined data.

Corpus	System	BLEU	NIST	TER	METEOR
TED	BASE	16.96	5.24	67.04	49.40
	MONO	10.66	4.13	74.63	41.02
	BI	11.90	4.13	74.59	42.46
BTEC	BASE	8.66	2.73	85.27	27.22
	MONO	8.46	2.71	84.45	27.14
	BI	8.50	2.74	83.84	27.30
EMEA	BASE	56.39	9.41	40.88	70.38
	MONO	13.72	3.95	89.58	39.23
	BI	14.07	4.05	89.12	40.22
EUP	BASE	25.74	6.54	58.08	48.46
	MONO	15.52	5.07	7155	51.01
	BI	16.61	5.24	71.08	52.49
OPEN	BASE	31.55	5.46	62.24	47.47
	MONO	9.90	3.08	84.02	32.88
	BI	10.67	3.21	83.12	34.35

Table 58. EN-to-PL translation results using bi-directional mined data.

Corpus	System	BLEU	NIST	TER	METEOR
TED	BASE	9.97	3.87	75.36	32.82
	MONO	6.90	3.09	81.21	27.00
	BI	7.14	3.18	78.83	27.76
BTEC	BASE	8.66	2.73	85.27	27.22
	MONO	8.46	2.71	84.45	27.14
	BI	8.76	2.78	82.30	27.39
EMEA	BASE	56.39	9.41	40.88	70.38
	MONO	13.66	3.95	77.82	32.16
	BI	13.64	3.93	77.47	32.83
EUP	BASE	25.74	6.54	58.08	48.46
	MONO	9.92	4.10	72.51	32.06
	BI	9.35	4.02	72.54	31.65
OPEN	BASE	31.55	5.46	62.24	47.47
	MONO	6.32	2.23	92.40	22.72
	BI	6.53	2.27	89.03	22.94

In addition, it was proven that mining data using two classifiers trained from a foreign to a native language and vice versa can significantly improve data quantity, even though some repetition is possible. Such bi-directional mining, which is logical, found additional data, mostly for wide domains. In narrow text domains, the potential gain is small. From a practical point of view, the method requires neither expensive training nor language-specific grammatical resources, but it produces satisfying results. It is possible to replicate such mining for any language pair or text domain, or for any reasonably comparable input data.

5.2.3 Improvements to the Yalign method

The results of experiments with methods for improving the performance of Yalign are shown in Table 59. It is important to note that these results are not comparable to those presented in Section 4.2.2, because the Yalign version used in the following experiments was additionally improved as described in Section 4.2.1. The data mining approaches shown in the tables are: Directional (PL->EN classifier) mining (MONO), bi-directional (additional EN->PL classifier) mining (BI), bi-directional mining with Yalign using a GPU-accelerated version of the Needleman-Wunsch algorithm (NW), and mining using a NW version of Yalign that was tuned (NWT). MONO is the total number of obtained parallel sentences used with each of classifiers presented in Table 53.

Table 59. Number of obtained bi-sentences.

Mining Method	Number of Bi-Sentences	Unique PL Tokens	Unique EN Tokens
MONO	510,128	362,071	361,039
BI	530,480	380,771	380,008
NW	1,729,061	595,064	574,542
NWT	2,984,880	794,478	764,542

As shown in Table 59, each of the improvements increased the number of parallel sentences discovered. In Table 60, the computational speed of the various versions of the Yalign algorithm are compared.

Table 60. Computation time of different Yalign versions.

Mining Method	Computation Time [s]
YALIGN	89.7
M YALIGN	14.7
NW YALIGN	17.3
GNW YALIGN	15.2

Tables 61 and 62 show the results of the data quality experiments using the following methods: The native Yalign implementation (Yalign), the multi-threaded implementation (M Yalign), Yalign with the Needleman-Wunsch algorithm (NW Yalign), and Yalign with a GPU-accelerated Needleman-Wunsch algorithm (GNW Yalign). In the tables, BASE represents the baseline system; MONO, the system enhanced with a mono-directional classifier; BI, a system with bi-directional mining; NW, a system mined bi-directionally using the Needleman-Wunsch algorithm; and TNW, a system with additionally tuned parameters.

The results indicate that multi-threading significantly improved speed, which is very important for large-scale mining. As anticipated, the Needleman-Wunsch algorithm decreases speed (which is why authors of the Yalign did not use it, in the first place). However, GPU acceleration makes it possible to obtain performance almost as fast as that of the multi-threaded A* version. It must be noted that the mining time may significantly differ when the alignment matrix is large (i.e., text is long). The experiments were conducted on a hyper-

Table 61. Results of SMT enhanced comparable corpora for PL-to-EN translation.

		BLEU	NIST	TER	METEOR
TED	BASE	16.96	5.26	67.10	49.42
	MONO	16.97	5.39	65.83	50.42
	BI	17.34	5.37	66.57	50.54
	NW	17.45	5.37	65.36	50.56
	TNW	17.50	5.41	64.36	90.62
EUP	BASE	36.73	8.38	47.10	70.94
	MONO	36.89	8.34	47.12	70.81
	BI	36.56	8.34	47.33	70.56
	NW	35.69	8.22	48.26	69.92
	TNW	35.26	8.15	48.76	69.58
EMEA	BASE	62.60	10.19	36.06	77.48
	MONO	62.48	10.20	36.29	77.48
	BI	62.62	10.22	35.89	77.61
	NW	62.69	10.27	35.49	77.85
	TNW	62.88	10.26	35.42	77.96
OPEN	BASE	64.58	9.47	33.74	76.71
	MONO	65.77	9.72	32.86	77.14
	BI	65.87	9.71	33.11	76.88
	NW	65.79	9.73	33.07	77.31
	TNW	65.91	9.78	32.22	77.36

Table 62. Results of SMT-enhanced comparable corpora for EN-to-PL translation.

		BLEU	NIST	TER	METEOR
TED	BASE	10.99	3.95	74.87	33.64
	MONO	11.24	4.06	73.97	34.28
	BI	11.54	4.02	73.75	34.43
	NW	11.59	3.98	74.49	33.97
	TNW	11.98	4.05	73.65	34.56
EUP	BASE	25.74	6.56	58.08	48.46
	MONO	24.71	6.37	59.41	47.45
	BI	24.63	6.35	59.73	46.98
	NW	24.13	6.29	60.23	46.78
	TNW	24.32	6.33	60.01	47.17
EMEA	BASE	55.63	9.31	41.91	69.65
	MONO	55.38	9.25	42.37	69.35
	BI	56.09	9.32	41.62	69.89
	NW	55.80	9.30	42.10	69.54
	TNW	56.39	9.41	40.88	70.38
OPEN	BASE	31.55	5.46	62.24	47.47
	MONO	31.27	5.45	62.43	47.28
	BI	31.23	5.40	62.70	47.03
	NW	31.47	5.46	62.32	47.39
	TNW	31.80	5.48	62.27	47.47

threaded Intel Core i7 CPU and a GeForce GTX 660 GPU. The quality of the data obtained with the NW algorithm version, as well as the TNW version, seems promising. Slight improvements in translation quality were observed, but more importantly, much more parallel data was obtained.

Statistical significance tests were conducted in order to evaluate how the improvements differ from each other. For each text domain, it was tested how significant the obtained results were from each other.

Changes with low significance were marked with "*", significant changes were marked with "**", and very significant with "***" in Tables 64 and 65. The significance test results in Table 63 showed visible differences in most row-to-row comparisons. In Table 64, all results were determined to be significant. Tables 65 and 66 show the results for an SMT system trained with only data extracted from comparable corpora (not the original, in-domain data). In the tables, BASE indicates the results for the baseline system trained on the original in-domain data; MONO, a system trained only on mined data in one direction; BI, a system trained on data mined in two directions with

Table 63. Significance of improvements in PL-EN translation.

		P-value
TED	MONO	0.0622*
	BI	0.0496**
	NW	0.0454**
	TNW	0.3450
EUP	MONO	0.3744
	BI	0.0302**
	NW	0.0032***
	TNW	0.0030***
EMEA	MONO	0.4193
	BI	0.0496**
	NW	0.0429**
	TNW	0.0186**
OPEN	MONO	0.0346**
	BI	0.0722*
	NW	0.0350**
	TNW	0.0259**

duplicate segments removed; NW, a system using bi-directionally mined data with the Needleman-Wunsch algorithm; and TNW, a system with additionally tuned parameters.

Table 64. Significance of improvements in EN-PL translation.

		P-value
TED	MONO	0.0305**
	BI	0.0296**
	NW	0.0099***
	TNW	0.0084***
EUP	MONO	0.0404**
	BI	0.0560*
	NW	0.0428**
	TNW	0.0623*
EMEA	MONO	0.0081***
	BI	0.0211**
	NW	0.0195**
	TNW	0.0075***
OPEN	MONO	0.0346**
	BI	0.0722*
	NW	0.0350**
	TNW	0.0259**

Significance tests were conducted in order to verify whether or not the results of Tables 65 and 66 were meaningful. For each text domain, the significance of the obtained results was compared with one another. Changes with low significance were marked with "*", significant changes were marked with "**", and very significant with "***" in Tables 67 and 68.

The results obtained in Tables 67 and 68 prove that results obtained in the experiment were highly significant.

The results presented in Tables 65 and 66 show a slight improvement in translation quality, which verifies that the improvements to Yalign positively impact the overall mining process. It must be noted that the above mining experiments were conducted using a classifier trained only on the TED data. This is why the improvements are very visible on this corpora and less visible on other corpora. In addition, similar to conclusions obtained in Chapter 4.2, improvements obtained by enriching the training set were observed mostly on wide-domain text, while for narrow domains the effects were negative. The improvements were mostly observed on the TED data set, because the

Table 65. SMT results using only comparable corpora for PL-to-EN translation.

		BLEU	NIST	TER	METEOR
TED	BASE	16.96	5.26	67.10	49.42
	MONO	12.91	4.57	71.50	44.01
	BI	12.90	4.58	71.13	43.99
	NW	13.28	4.62	71.96	44.47
	TNW	13.94	4.68	71.50	45.07
EUP	BASE	36.73	8.38	47.10	70.94
	MONO	21.82	6.09	62.85	56.40
	BI	21.24	6.03	63.27	55.88
	NW	20.20	5.88	64.24	54.38
	TNW	20.42	5.88	63.95	54.65
EMEA	BASE	62.60	10.19	36.06	77.48
	MONO	21.71	5.09	74.30	44.22
	BI	21.45	5.06	73.74	44.01
	NW	21.47	5.06	73.81	44.14
	TNW	22.64	5.30	72.98	45.52
OPEN	BASE	64.58	9.47	33.74	76.71
	MONO	11.53	3.34	78.06	34.71
	BI	11.64	3.25	82.38	33.88
	NW	11.64	3.32	81.48	34.62
	TNW	11.85	3.35	77.59	35.08

Table 66. SMT results using only comparable corpora for EN-to-PL translation.

		BLEU	NIST	TER	METEOR
TED	BASE	10.99	3.95	74.87	33.64
	MONO	7.89	3.22	83.90	29.20
	BI	7.98	3.27	84.22	29.09
	NW	8.50	3.28	83.02	29.88
	TNW	9.15	3.38	78.75	30.08
EUP	BASE	25.74	6.56	58.08	48.46
	MONO	14.48	4.76	70.64	36.62
	BI	13.91	4.67	71.32	35.83
	NW	13.13	4.54	72.14	35.03
	TNW	13.41	4.58	71.83	35.34
EMEA	BASE	56.39	9.41	40.88	70.38
	MONO	19.11	4.73	74.77	37.34
	BI	18.65	4.60	75.16	36.91
	NW	18.60	4.53	76.19	36.30
	TNW	18.58	4.48	76.60	36.28
OPEN	BASE	31.55	5.46	62.24	47.47
	MONO	7.95	2.40	88.55	24.37
	BI	8.20	2.40	89.51	24.49
	NW	9.02	2.52	86.14	25.01
	TNW	9.20	2.52	84.86	25.20

Table 67. Significance test for PL-EN translation.

		P-value
TED	MONO	0.0167**
	BI	0.0209**
	NW	0.0177**
	TNW	0.0027***
EUP	MONO	0.000***
	BI	0.000***
	NW	0.000***
	TNW	0.000***
EMEA	MONO	0.000***
	BI	0.000***
	NW	0.000***
	TNW	0.000***
OPEN	MONO	0.000***
	BI	0.000***
	NW	0.000***
	TNW	0.000***

Table 68. Significance test for EN-PL translation.

		P-value
TED	MONO	0.0215**
	BI	0.0263**
	NW	0.0265**
	TNW	0.0072***
EUP	MONO	0.000***
	BI	0.000***
	NW	0.000***
	TNW	0.000***
EMEA	MONO	0.0186**
	BI	0.0134**
	NW	0.0224**
	TNW	0.0224**
OPEN	MONO	0.000***
	BI	0.000***
	NW	0.000***
	TNW	0.000***

classifier was only trained on text samples. Regardless, the text domain tuning algorithm proved to always improve translation quality.

With the help of techniques explained earlier, we were capable of creating comparable corpora for many PL-* language pairs and, later, probe them for parallel phrases. We paired Polish (PL) with Arabic (AR), Czech (CS), German (DE), Greek (EL), English (EN), Spanish (ES), Persian (FA), French (FR), Hebrew (HE), Hungarian (HU), Italian (IT), Dutch (NL), Portuguese (PT), Romanian (RO), Russian (RU), Slovenian (SL), Turkish (TR) and Vietnamese (VI). Statistics of the resulting corpora are presented in Table 69.

In order to assess the corpora quality and usefulness, we trained the baseline SMT systems by utilizing the WIT[38] data (BASE). We also augmented them with resulting mined corpora both as parallel data as well as the language models (EXT). The additional corpora were domain-adapted through the linear interpolation and Modified Moore-Lewis filtering [250]. Tuning of the system was not executed during experiments due to the volatility of the MERT [21]. However, usage of the MERT would have an overall positive impact on MT system in general [21]. The results are showed in Table 70.

---

[38] https://wit3.fbk.eu/mt.php?release=2013-01

Table 69. Results of mining after progress.

Language Pair	Number of bi-sentences	Number of unique PL tokens	Number of unique foreign tokens
PL-AR	823,715	1,345,702	1,541,766
PL-CS	62,507	197,499	206,265
PL-DE	169,739	345,266	341,284
PL-EL	12,222	51,992	51,384
PL-EN	172,663	487,999	412,759
PL-ES	151,173	411,800	377,557
PL-FA	6,092	31,118	29,218
PL-FR	51,725	215,116	206,621
PL-HE	10,006	42,221	47,645
PL-HU	41,116	130,516	136,869
PL-IT	210,435	553,817	536,459
PL-NL	167,081	446,748	425,487
PL-PT	208,756	513,162	491,855
PL-RO	6,742	38,174	37,804
PL-RU	170,227	365,062	440,520
PL-SL	17,228	71,572	71,469
PL-TR	15,993	93,695	92,439
PL-VI	90,428	240,630	204,464

The assessment was based on sets of official test sets from IWSLT 2013[39] conference. Bilingual Evaluation Understudy (BLEU) measurement was used to score the progress. As it was expected earlier, sets of supplementary data enhance the general quality of translation for each and every language.

In order to verify the importance of our results, we conducted the significance tests for 4 diverse languages. The decision was made to use the Wilcoxon test. The Wilcoxon test (also known as the signed-rank test or the matched-pairs test) is one of the most popular alternatives for the Student's t-test for dependent samples. It belongs to the group of non-parametric tests. It is used to compare two (and only two) dependent groups, that is, two measurement variables. The significance tests were conducted in order to evaluate how the improvements differ from each other. Changes with low significance were marked with *, significant changes were marked with ** and very significant with *** in presented Tables 71 and 72.

---

[39] Iwslt.org

Table 70. Results of MT experiments.

LANGUAGE	SYSTEM	DIRECTION	BLEU	LANGUAGE	SYSTEM	DIRECTION	BLEU	LANGUAGE	SYSTEM	DIRECTION	BLEU
PL-AR	BASE	→PL	19.67	PL-FA	BASE	→PL	14.21	PL-PT	BASE	→PL	27.07
	EXT	→PL	21.78		EXT	→PL	14.32		EXT	→PL	29.14
	BASE	←PL	20.98		BASE	←PL	16.87		BASE	←PL	30.11
	EXT	←PL	23.12		EXT	←PL	17.03		EXT	←PL	31.33
PL-CS	BASE	→PL	12.21	PL-FR	BASE	→PL	19.07	PL-RO	BASE	→PL	22.16
	EXT	→PL	12.98		EXT	→PL	20.01		EXT	→PL	22.26
	BASE	←PL	13.44		BASE	←PL	21.13		BASE	←PL	25.01
	EXT	←PL	14.21		EXT	←PL	21.56		EXT	←PL	25.67
PL-DE	BASE	→PL	23.68	PL-HE	BASE	→PL	17.03	PL-RU	BASE	→PL	12.36
	EXT	→PL	24.91		EXT	→PL	17.65		EXT	→PL	13.51
	BASE	←PL	26.61		BASE	←PL	18.18		BASE	←PL	13.58
	EXT	←PL	26.87		EXT	←PL	18.54		EXT	←PL	14.32
PL-EL	BASE	→PL	14.27	PL-HU	BASE	→PL	14.62	PL-SL	BASE	→PL	12.11
	EXT	→PL	14.67		EXT	→PL	15.23		EXT	→PL	12.57
	BASE	←PL	17.22		BASE	←PL	17.18		BASE	←PL	14.26
	EXT	←PL	17.28		EXT	←PL	17.81		EXT	←PL	14.61
PL-EN	BASE	→PL	15.91	PL-IT	BASE	→PL	18.83	PL-TR	BASE	→PL	11.59
	EXT	→PL	17.01		EXT	→PL	19.87		EXT	→PL	12.68
	BASE	←PL	17.09		BASE	←PL	21.19		BASE	←PL	13.07
	EXT	←PL	18.43		EXT	←PL	21.34		EXT	←PL	13.44
PL-ES	BASE	→PL	16.35	PL-NL	BASE	→PL	18.29	PL-VI	BASE	→PL	12.66
	EXT	→PL	17.92		EXT	→PL	20.13		EXT	→PL	14.12
	BASE	←PL	18.34		BASE	←PL	20.79		BASE	←PL	14.11
	EXT	←PL	18.65		EXT	←PL	21.45		EXT	←PL	15.17

Table 71. Significance tests pl->*.

Language Pair	Number of bi-sentences
PL-EN	0.0103**
PL-CS	0.0217**
PL-AR	0.0011***
PL-VI	0.0023***

Table 72. Significance tests * ->PL.

Language Pair	Number of bi-sentences
PL-EN	0.0193**
PL-CS	0.0153**
PL-AR	0.0016***
PL-VI	0.0027***

Bi-lingual sentence extraction has particular importance in dealing with unsubstantiated learning processes for multiple tasks involved. With the help of this methodology, we can easily resolve the dissimilarities between Polish and other languages. It is a method that is independent from language matters, it is adaptable to new environments for any language pair. Our experiments validated the performance of the method. The corpora received as consequence of the experiments can maximize the quality of MT in an under-resourced text domain. However, in few scenarios, only small differences have been observed in BLEU scores. Keeping that aside, it can be said that even such small differences, can influence the real life situations positively especially for infrequent translation cases. Moreover, the results of our work are freely accessible for the research community (corpora is hosted at OPUS[40] and tools at GitHub[41]). In order to see things from a sensible outlook, we can say that such methodology does not require large scale training or special language specific rammer resources, and inspite of that they produce gratifying results.

Because statistically classified data contains some amounts of noisy data, in future we plan to develop precise filtering strategies for bi-lingual corpora. The results of current solution are highly related to SVM classifier. In other words, we plan to train more classifiers for different text domains in order to discover more bi-lingual sentences.

---

[40] http://opus.lingfil.uu.se/Wikipedia.php
[41] https://github.com/krzwolk/yalign

5.2.4 *Parallel data mining using pipeline of tools*

Sentence alignment by human translators was compared to the tool pipeline for mining parallel data in experiments. Information about the human alignment is presented in Table 73.

Table 73. Human alignment.

No.	Vocab.Count		Sentences		Human Aligned	No.	Vocab.Count		Sentences		Human Aligned
	PL	EN	PL	EN			PL	EN	PL	EN	
1	2526	1910	324	186	127	11	2861	2064	412	207	8
2	2950	3664	417	596	6	12	2186	1652	345	188	2
3	504	439	45	43	34	13	2799	3418	496	472	124
4	2529	1383	352	218	4	14	1164	1037	184	196	3
5	807	1666	104	275	10	15	2465	1781	365	189	3
6	2461	4667	368	618	1	16	1946	1839	282	198	132
7	2701	1374	560	210	16	17	966	782	113	96	7
8	1647	768	274	78	1	18	2005	1253	309	134	1
9	1189	1247	121	120	64	19	2443	2001	251	189	21
10	2933	3047	296	400	150	20	9888	3181	1297	367	2

In the "Vocab Count" column, the number of distinct words and their forms are presented, in "Sentences," the number of recognized sentences in each language, and in "Human Aligned," the number of sentence pairs aligned by a human.

The same articles were processed with the described pipeline.

Table 74. Automatic alignment.

No.	Hunaligned		Filtered		No.	Hunaligned		Filtered	
	YES	NO	YES	NO		YES	NO	YES	NO
1	109	130	18	0	11	8	325	0	0
2	6	527	25	2	12	2	256	0	0
3	17	24	0	0	13	70	414	1	0
4	4	302	1	0	14	3	182	0	0
5	6	211	1	0	15	3	285	0	0
6	1	498	0	0	16	111	108	0	0
7	16	440	0	0	17	7	98	0	0
8	1	221	0	0	18	1	202	0	0
9	51	62	0	0	19	21	192	0	0
10	127	245	0	0	20	2	1078	0	0

Table 74 shows how many sentences Hunalign initially aligned as similar and how many of them remained after filtering with the tool described in Chapter 4.7.

Both the columns "YES" and "NO" under the Hunaligned heading are aligned sentences. The numbers represent how many of them were aligned correctly and how many had alignment errors. In the "Filtered" column, the number of parallel sentences that remained after filtering is shown. In the "YES" column, properly-aligned sentences are indicated, and in the "NO" column, the erroneously-aligned sentences are given. In this experiment, human translators were asked to check which of the remaining sentence pairs were truly parallel and if any pairs were missed.

The author's new method for obtaining, mining and filtering very parallel, noisy-parallel and comparable corpora was also introduced. The experiments show that the automated methods provide good accuracy and some correlation with human judgment. Nevertheless, there is still room for improvement in two areas. In the experiments, the amount of data obtained via automated methods in comparison with human work is not satisfactory. First, Hunalign would have performed much better if it was provided a good quality dictionary, especially one that contains in-domain vocabulary. Second, a statistical machine translation system would have greatly increased quality by providing better translations. After the initial mining of the corpora, the parallel data obtained can potentially be used for both purposes. First, a phrase-table can be trained from extracted bi-sentences, and it would be easy to extract a good in-domain dictionary (also including probabilities of translations) from the table. Second, the SMT can be retrained with newly-obtained bilingual sentence pairs from the mined data and adapted based on it [84]. Lastly, the pipeline can be re-run with the new capabilities. The steps can be repeated until the extraction results are fully satisfactory.

5.2.5 *Analogy-based method*

The results presented in Table 75 were obtained using a corpus of sentences generated with the analogy detection method. Expanding the corpus with newly-generated sentences gave decreasing results for all metrics.

Table 75. Results on TED corpus trained with additional analogy-based corpus.

PL-EN	BLEU	NIST	TER	MET
TED Baseline	19.69	5.24	67.04	49.40
Analogy corpus	16.44	5.15	68.05	49.02
EN-PL	BLEU	NIST	TER	MET
TED Baseline	9.97	3.87	75.36	32.82
Analogy corpus	9.74	3.84	75.21	32.55

The most likely reason for such results is that the analogy method is designed to extend existing parallel corpora from available non-parallel data. However, in order to establish a meaningful baseline, it was decided to test a noisy-parallel corpus independently mined using this method. Therefore, the results are less favorable than those obtained using a Yalign-based method. It seems that the problem is that the analogy-based method, in contrast with Yalign-based methods, does not mine domain-specific data.

Additionally, after manual data analysis, it was noticed that the analogy-based method suffers from duplicates and a relatively large amount of noisy data, or low-order phrases. As a solution to this problem, it was decided to apply two different methods of filtering. The first one is simple, based on length of the sentences in a corpus. In addition, all duplicates and very short (less than 10 characters) sentences were removed. As a result, it was possible to retain 58,590 sentences in the final corpus. The results are denoted in Table 76 as FL1. Second, the filtration method described in Section 4.7.2 was used (FL2). The number of unique EN tokens before filtration was 137,262, and there were 139,408 unique PL tokens. After filtration, there were 28,054 and 22,084 unique EN and PL tokens, respectively.

Table 76. Filtration results for analogy-based method (number of bi-sentences).

Number of sentences in base corpus	3,800,000
Number of rewriting models	8,128
Bi-sentences in base corpus	114,107
Bi-sentences after duplicates removal	64,080
Remaining bi-sentences after filtration (FL1)	58,590
Remaining bi-sentences after filtration (FL2)	6,557

5.3 Quasi comparable corpora exploration

The aim of this part of research was the preparation of parallel and quasi-comparable corpora and language models. This work improves SMT quality through the processing and filtering of parallel corpora and through extraction of additional data from the resulting quasi-comparable corpora. To enrich the language resources of SMT systems, adaptation and interpolation techniques will be applied to the prepared data. Experiments were conducted using data from a wide domain (TED[42] presentations on various topics).

Evaluation of SMT systems was performed on random samples of parallel data using automated algorithms to evaluate the quality and potential usability of the SMT systems' output [22].

---

[42] https://www.ted.com/

In this part, methodologies that obtain parallel corpora from data sources that are not sentence-aligned and very non-parallel, such as quasi-comparable corpora, are presented. The results of initial experiments on text samples obtained from the Internet crawled data are presented. The quality of the approach used was measured by improvements in MT systems translations.

For the experiments in data mining, the TED corpora prepared for the IWSLT 2014 evaluation campaign by the FBK[43] was chosen. This domain is very wide and covers many unrelated subject areas. The data contains almost 2.5 M untokenized words [15]. The experiments were conducted on PL-EN (Polish-English) corpora.

The solution can be divided into three main steps. First, the quasi-comparable data is collected, then it is aligned based on keywords, and finally the aligned results are mined for parallel sentences. The last two steps are not trivial, because there are great disparities between polish-english documents. Text samples in English corpus are mostly misaligned, with translation lines whose placement does not correspond to any text lines in the source language. Moreover, most sentences have no corresponding translations in the corpus at all. The corpus might also contain poor or indirect translations, making alignment difficult.

Thus, alignment is crucial for accuracy. Sentence alignment must also be computationally feasible to be of practical use in various applications.

Before a mining tool processes the data, it must be prepared. Firstly, all the data is downloaded and saved in a database. In order to obtain data, a web crawler was developed. As input, our tool requires bi-lingual dictionary or a phrase table with probability of such translation. Such input can be obtained using parallel corpora and tools like GIZA++ [148]. Based on such bi-lingual translation equivalents it is possible to make a query to the Google Search engine. Secondly, we applied filtering of not likely translations and limiting number of crawled search results. In this research we used only 1-gram dictionary of words that for 70% were translations of each other. For each keyword, we crawled only 5 pages. Such strict limits were necessary because of the time required for web crawling, however, it is obvious that much more data with better precision and domain adaptation will be obtained when crawling higher order n-grams. In summary, 43,899 pairs were used from the dictionary, which produced almost 3.4 GB of data that contained 45,035,931 lines in english texts and 16,492,246 in polish texts. The average length of EN articles was equal to 3,724,007 tokens and of PL to 4,855,009.

Secondly, our tool aligns article pairs and unifies the encoding of articles that do not exist in UTF-8. These keyword-aligned articles are filtered to

---

[43] http://www.fbk.eu/

remove any HTML tags, XML tags or noisy data (tables, references, figures, etc.). Finally, bilingual documents are tagged with a unique ID as a quasi-comparable corpus. To extract the parallel sentence pairs, a decision was made to try a strategy designed to automate the parallel text mining process by finding sentences that are close translation matches from quasi-comparable corpus. This presents opportunities for harvesting parallel corpora from sources, like translated documents and the web, that are not limited to a particular language pair. However, alignment models for two selected languages must first be created. For this, the same methodology as for the comparable corpora was used.

As already mentioned, some methods for improving the performance of the native classifier were developed. First, speed improvements were made by introducing multi-threading to the algorithm, using a database instead of plain text files or Internet links, and using GPU acceleration in sequence comparison. More importantly, two improvements were made to the quality and quantity of the mined data. The A* search algorithm was modified to use Needleman-Wunsch, and a tuning script of mining parameters was developed. In this section, the PL-EN TED corpus will be used to demonstrate the impact of the improvements (it was the only classifier used in the mining phase). The data mining approaches used were: Directional (PL->EN classifier) mining (MONO), bi-directional (additional EN->PL classifier) mining (BI), bi-directional mining using a GPU-accelerated version of the Needleman-Wunsch algorithm (NW), and mining using the NW version of the classifier that was tuned (NWT). The results of such mining are shown in Table 77.

Table 77. Obtained bi-sentences.

Mining Method	Number of Bi-Sentences
MONO	21,132
BI	23,480
NW	24,731
NWT	27,723

Table 78. Computation time.

Mining Method	Computation Time [s]
Y	272.27
MY	63.1
NWMY	112.6
GNWMY	74.4

As presented in Table 78, each of the improvements increased the number of parallel sentences discovered. In addition, in Table 78 a speed comparison is made using different versions of the tool. A total of 1,000 comparable articles were randomly selected from Wikipedia and aligned using the native implementation (Y), multi-threaded implementation (MY), classifier with the Needleman-Wunsch algorithm (NWMY), and with a GPU-accelerated Needleman-Wunsch algorithm (GNWMY).

The results indicate that multi-threading significantly improved speed, which is very important for large-scale mining. As anticipated, the Needleman-Wunsch algorithm decreases speed. However, GPU acceleration makes it possible to obtain performance almost as fast as that of the multi-threaded A* version. It must be noted that the mining time may significantly differ when the alignment matrix is big (text is long). The experiments were conducted on a hyper-threaded Intel Core i7 CPU and a GeForce GTX 980 GPU.

Using techniques described above, we were able to build quasi-comparable corpora and mine it for parallel sentences for the Polish-English language pair and evaluate it using data being part of IWSLT 2014 conference. The obtained parallel corpus (EXT) and the TED corpora statistics are presented in Table 79.

Table 79. Corpora statistics.

Corpus	Number of bi-sentences	Number of unique EN tokens	Number of unique foreign tokens
TED	2,459,662	2,576,938	2,864,554
EXT	818,300	1,290,000	1,120,166

Table 80. Results of MT experiments.

LANG	SYSTEM	DIRECTION	BLEU
PL-EN	BASE	→EN	30.21
	EXT	→EN	31.37
	BASE	EN←	21.07
	EXT	EN←	22.47

To evaluate the corpora, we trained baseline systems using IWSLT 2014 official data sets and enriched them with obtained quasi-comparable corpora, both as parallel data and as language models. The enriched systems were trained with the baseline settings but additional data was adapted using linear interpolation and Modified Moore-Levis [82]. Because of the well know MERT instability, tuning was not performed in the experiments [21].

The evaluation was conducted using official test sets from IWSLT 2010-2013 campaigns and averaged. For scoring purposes, Bilingual Evaluation

Understudy (BLEU) metric was used. The results of the experiments are shown in Table 80. BASE in Table 80 stands for baseline system and EXT for enriched systems.

As anticipated, additional data sets improved overall translation quality for each language and in both translation directions. The gain in quality was observed mostly in the English to foreign language direction.

5.4 Other fields of MT techniques application

The aim of this chapter is to present current methods in MT used in other field of research and also to show how data pre-processing in certain situations can improve translation quality.

5.4.1 *HMEANT and other metrics in re-speaking quality assessment*

Re-speaking is a mechanism for obtaining high quality subtitles for use in live broadcast and other public events. Because it relies on humans performing the actual re-speaking, the task of estimating the quality of the results is non-trivial. Most organisations rely on humans to perform the actual quality assessment, but purely automatic methods have been developed for other similar problems, like Machine Translation. This research will try to compare several of these methods: BLEU, EBLEU, NIST, METEOR, METEOR-PL, TER, RIBES and HMEANT. These will then be matched to the human-derived NER metric, commonly used in re-speaking.

One of the main driving forces in Speech Technology, for the last several years, comes from the efforts of various groups and organizations tackling with the issue of disability, specifically deaf and hearing-impaired people. Most notably, a long term effort by such organisations has lead to a plan by the European Commision to enable "Subtitling of 100% of programs in public TV all over the EU by 2020 with simple technical standards and consumer friendly rules" [251]. This ambitious task would not be possible to achieve without the aid of Speech Technology. While there has been a considerable improvement of quality of Automatic Speech Recognition (ASR) technology recently, many of the tasks present in real-life are simply beyond complete automation. On the other hand, there are tasks, which are also impossible to achieve by humans without the aid of ASR. For example, movie subtitles are usually done by human transcribers and can take a day, up to a week, per material to complete. Live subtitling, however, can sustain only a few seconds delay between the time an event is recorded and the time it appears on the viewer's screen. This is where re-speaking comes into play. The idea of re-speaking is to use ASR to create live and fully annotated subtitles, but rather than risking misrecognition of the speech happening in the live recording, a specially trained individual, the so-called re-speaker, repeats the speech from the recorded event in a quiet and controlled environment. This approach has many advantages that guarantee excellent results: Controlled environment, the ability to adapt the speaker, the ability of the speaker to adapt to the software, solving problems like double-speak, cocktail party effect and non-grammatic speech by paraphrasing. The standard of quality required by many jurisdictions demands less than 2% of errors [252]. From the point of view of a re-speaker, this problem is very similar to that of an interpreter only instead of translating from one language to another, re-speaking is usually done within one language only. There are many aspects of re-speaking worthy of their own insight [253], but this monograph will deal only with the issue of quality assessment. Similarly to Machine Translation (MT), the assessment of the accuracy of re-speaking is not a trivial task, because there are many possible ways to paraphrase an utterance, just like there are many ways to translate any given sentence. Measuring the accuracy of such data has to take semantic meaning into account, rather than blindly performing simple word-to-word comparison. One option is to use humans to perform this evaluation, as in the NER model, described later in the monograph. This has been recognized as very expensive and time-consuming [254]. As a result, human effort cannot keep up with the growing and continual need for the evaluation. This led to the recognition that the development of automated evaluation techniques is critical [254, 26]. Unfortunately most of the automatic evaluation metrics were developed for other purposes than re-speaking (mostly machine translation) and are not suited for languages like Polish that differ semantically and structurally from English. Polish has complex declension, 7 cases, 15 gender forms and complicated grammatical construction. This leads to a larger vocabulary and greater complexity in data requirements for such tasks. Unlike Polish, English does not have declensions. In addition, word order, especially the Subject-Verb-Object (SVO) pattern, is absolutely crucial in determining the meaning of an English sentence [18]. While these automatic metrics have already been thoroughly studied, we feel that there is still much to be learnt, especially in different languages and for different tasks, like re-speaking. This monograph will compare some of these automated and human assisted metrics, while also considering issues related specifically to Polish. A small introduction to all the metrics is presented in the beginning of the monograph, followed by an experiment using actual re-speaking data.

5.4.1.1 RIBES metric

The focus of the RIBES metric is word order. It uses rank correlation coefficients based on word order in order to compare SMT and reference translations. The primary rank correlation coefficients used are Spearman's ρ, which measures the distance of differences in rank, and Kendall's τ, which measures the direction of differences in rank [18]. These rank measures can be normalized in order to ensure positive values [18]:

Normalized Spearman's ρ (STB) = (U + 1)/2

Normalized Kendall's X (SYA) = (X + 1)/2

These measures can be combined with precision P and modified to avoid overestimating the correlation of only corresponding words in the SMT and reference translations:

NSR Pα and NKT Pα

where α is a parameter in the range $0 < α < 1$.

5.4.1.2 NER metric

The NER model [255] is a simple extension of the word accuracy metric adapted specifically for measuring the accuracy of subtitles. It is one of two measures that is of particular importance for media providers (television companies, movie distributors, etc.), the other one being the reduction rate. Generally, the aim of good subtitles is to reduce the length of written text as much as possible (in order to preserve space on screen and make it easier to read) while maintaining an almost perfect accuracy (usually above 98%).

Since we are dealing with paraphrasing, it is very difficult to perform accurate measurements by comparing the text only. The NER model gets around this problem by counting errors using a simple formula, which inspired its name:

$$\text{NER accuracy} = \frac{N - E - R}{N} \times 100\%$$

where N is the number of analysed tokens (usually also includes punctuation), E is the number of errors performed by the re-speaker, and R is the number of errors performed by the ASR system (on re-speaker's speech). Additionally, the errors in E are weighted: 0.25 for minor errors, 0.5 for normal and 1 for serious errors. There are user-friendly tools available for making these assessments and, obviously, there may be a certain level of human bias involved in these estimates. Nevertheless, all the decisions are thoroughly checked and explainable using this method, which makes it one of the most popular techniques for subtitle quality assessment used by many regulatory bodies worldwide.

5.4.1.3　*Data and experimental setup*

The data used in the experiments described in this monograph was collected during a study performed in the Institute of Applied Linguistics at the University of Warsaw [255]. This, still ongoing, study aims to determine the relevant features of good re-speakers and their technique. One of the obvious measures is, naturally, the quality of re-speaking discussed in this monograph.

The subjects were studied in several hour sessions, where they had to perform various tasks and psychological exams. The final test was to do actual re-speaking of pre-recorded material. This was simultaneously recorded and recognized by a commercial, off-the-shelf ASR suite. The software was moderately adapted to the re-speaker during a several hour session, a few weeks before the test.

The materials included four different 5 minute segments in the speaker's native language and one in English (where the task was also translation). The recordings were additionally transcribed by a human, to convey exactly what the person said. The final dataset contains three sets of transcriptions:

1. The transcription of the original recorded material
2. The transcription of the re-speaker transcribed by a human (used in the evaluation)
3. The output of the ASR recognizing the re-speaker

Table 81 contains results of human evaluation using NER and Reduction (the goal is to have as big NER and Reduction as possible at the same time) and semi-automatic using HMEANT metric.

Two annotators with no linguistic background were asked to participate. Each annotator evaluated exactly 15 re-spoken samples, each of which contained 40 sentences that were supposed to be annotated by them.

5.4.1.4　*Results*

Backward linear regression has been used in analysis for the following reasons:

- The data is linear (correlation analysis present linear relation)
- The data is ratio level data, therefore, good for linear regression
- The regression analysis provides more than one regression result, thus, the best variables can be found.

Reliable variables are extracted by excluding irrelevant variables from the 1st to the last stage (give the most reliable variables).

For this analysis, the Standardized Coefficients would have the strength of relationship extracted from the Unstandardized Coefficients. The sig

Table 81. Results of human evaluation.

NER	HMEANT	REDUCTION
93,61	0,6016	22,77
94,09	0,6626	14,33
94,78	0,871	9,31
95,96	0,9395	2,71
94,77	0,6854	10,15
97,01	0,8958	3,89
95,83	0,7407	8,63
87,56	0,4621	25,21
85,76	0,4528	28,09
93,98	0,7148	8,63
95,79	0,5994	11,68
94,77	0,4641	19,12
97,41	0,9253	0,34
93,76	0,9529	5,75
93,89	0,8086	4,4
93,97	0,8194	7,45
92,74	0,725	8,46
88,07	0,6437	25,21
86,68	0,2976	32,66
84,36	0,4148	52,12
95,01	0,719	11,84
92,23	0,6244	11,84
93,11	0,5676	27,58
95,36	0,796	6,09
89,97	0,589	21,15
94,95	0,7975	5,41
95,75	0,8302	7,45
98,69	0,8869	4,23
89,37	0,5852	-6,6
93,73	0,8144	2,2

(p-value) was judged with the alpha (0.05) in order to investigate the most significant variables that explain the NER metrics. The Adjusted R-square (R^2) would provide the information, how much the variables are explaining the NER variances.

Table 82 represents the regression summary for NER with HMEANT and REDUCTION metrics. Here, at first, the model has both HMEANT and REDUCTION metrics and HMEANT are the significant predictors of NER. Additionally, REDUCTION is the insignificant metric, thus, it has been removed for the second model. In the second model, only HMEANT (p = 0.000) is significant as it has p-value below the alpha (0.05). In this case, the final model (2nd model) shows that the B-value for HMEANT is 0.162, indicating that a one-unit increase in HMEANT value would bring a 0.162 times increase in NER value, higher beta value (B) indicates strong relationship. The constant value in this case is equal to 81.824 (higher constant value indicates more error in the model). The Regression overall explains 59.4% variability of NER values, which is statistically acceptable.

Table 82. Regression result summary for NER and HMEANT, REDUCTION metrics.

Model		Unstandardized Coefficients		Standardized Coefficients	T	Sig.	Adjusted R-square
		B	Std. Error	Beta			
1st model	(Constant)	81.814	1.825		44.838	.000	0.603
	HMEANT	.156	.026	.744	5.925	.000	
	REDUCTION	.030	.038	.102	.810	.425	
2nd model	(Constant)	81.824	1.813		45.123	.000	0.594
	HMEANT	.162	.025	.770	6.395	.000	

a. Dependent Variable: NER

From the two models, it has been found that HMEANT is a significant predictor of NER (which means it correlates with human judgments correctly). The regression equation that can compute the value of NER based on HMEANT statistical metrics as:

$$NER = 81.82 + 0.162 * HMEANT$$

Moving forward, regression residual plots from the above regression for the significant metrics are presented in Figure 35. The histogram and normality plots show the distribution of residual, and the scatter plot shows relation between dependent metric with regression residual. The closer the dots in the plot to the regression line, the better the R-square value and the better the relationship.

Figure 35, shows the regression residual analysis for HMEANT, and it is clear that the histogram and normality plot show that the residuals are distributed normally.

Lastly, we analysed the IAA, dividing it into annotation and alignment steps. The agreement in annotation was quite high, equal to 0.87, but in the alignment step the agreement was 0.63, most likely because of the subjective human feelings about the meaning. In addition, we measured that the average

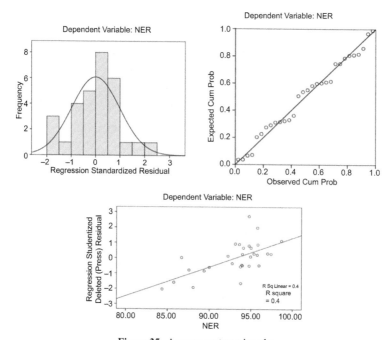

Figure 35. Average sentence lengths.

time required for an inexperienced annotator to evaluate one 40 sentence long text was about 27 minutes.

As far as other automatic metrics are concerned, we used 10 transcripts prepared using the protocol above. Each transcript was evaluated with all the metrics described in this monograph as well as manually using the NER metric. Table 83 presents evaluation between human made text transcription and the original texts. Table 84 presents evaluation between ASR system and original texts.

Table 85 represents the regression summary for NER and other metrics. Here, at first, the model has all the metrics and, except EBLEU, none of them are significant predictors of NER. Additionally, among all the metrics, TER is the most insignificant metric, therefore, it has been removed for the second model. In the second model, only EBLEU (p = 0.007) and NIST (p = 0.019) were significant, as they were below the alpha (0.05). Here in the 2nd model, the most insignificant metrics is RIBES (p = 0.538), therefore, it was removed in the 3rd model at the next stage. In the 3rd model, again EBLEU (p = 0.006) and NIST (p = 0.018) are significant and their significance value increased so did the Beta for EBLEU. The higher those values are the more accurate they become in predicting NER. In this model, METEOR_pl is the most insignificant metric, therefore, it has been removed for the next model.

Table 83. Evaluation of human made text transcription and original text (RED—reduction rate).

SPKR	BLEU	NIST	TER	METEOR	METEOR-PL	EBLEU	RIBES	NER	RED
1	56.20	6.90	29.10	79.62	67.07	63.32	85.59	92.38	10.15
2	56.58	6.42	30.96	78.38	67.44	58.82	86.13	94.86	17.77
3	71.56	7.86	18.27	88.28	76.19	79.58	92.48	94.71	12.01
4	76.64	8.27	13.03	90.34	79.29	87.38	93.07	93.1	3.72
5	34.95	5.32	44.50	61.86	47.74	37.06	71.60	91.41	17.03
6	61.73	7.53	20.47	83.10	72.55	69.11	92.43	92.33	4.89
7	61.74	6.93	28.26	78.26	69.78	63.99	78.32	95.3	10.29
8	33.52	4.28	46.02	63.06	47.61	36.75	77.55	93.95	26.81
9	68.97	7.46	22.50	83.15	76.56	71.83	88.78	94.73	4.05
10	70.02	7.80	18.78	86.12	78.71	75.16	88.15	95.23	6.41
11	47.07	5.56	33.84	76.10	62.02	47.86	85.05	93.61	22.77
12	53.49	6.65	30.63	77.93	65.27	55.90	86.20	94.09	14.33
13	75.71	7.95	16.07	89.72	83.13	77.74	91.62	94.78	9.31
14	66.46	7.60	18.44	84.34	76.20	66.53	90.76	94.82	6.09
15	25.77	1.85	54.65	58.62	40.82	31.08	68.90	85.26	32.83
16	88.82	8.66	5.75	96.15	90.12	95.20	95.76	95.96	2.71
17	63.26	7.25	25.72	81.95	73.16	65.83	88.99	94.77	10.15
18	60.69	7.18	26.23	79.41	70.86	66.77	87.56	95.15	5.75
19	59.13	7.2	25.04	80.00	70.32	62.77	89.17	95.78	4.74
20	86.24	8.43	7.11	94.60	90.07	92.88	95.39	95.58	1.52
21	20.61	2.08	65.14	47.56	33.31	27.42	55.63	91.62	36.89
22	64.40	7.43	22.84	82.69	72.82	66.93	89.44	93.46	10.15
23	27.30	2.69	52.62	58.96	43.19	35.78	68.06	90.07	31.81
24	82.43	8.33	10,15	92.40	86.04	85.61	94.63	97.18	12.02
25	82.22	8.44	9.48	93.75	87.00	91.59	95.12	97.01	3.89
26	76.17	8.25	13.54	91.33	81.54	82.35	91.95	95.83	8.63
27	35.01	4.39	49.41	62.64	47.81	46.64	72.79	87.56	25.21
28	29.50	3.53	53.13	60.73	42.98	40.87	74.25	85.76	28.09
29	70.04	7.78	17.26	87.22	80.63	73.58	91.56	93.98	8.63
30	56.75	6.89	26.06	79.60	70.52	58.43	90.78	95.79	11.68
31	63.18	6.90	26.57	83.47	71.80	67.80	84.51	94.29	17.77
32	31.74	5.14	43.15	63.58	49.37	33.07	83.05	94.77	19.12

Table 83 cont. ...

...Table 83 cont.

SPKR	BLEU	NIST	TER	METEOR	METEOR-PL	EBLEU	RIBES	NER	RED
33	89.09	8.54	5.41	95.69	91.62	92.62	95.94	97.41	0.34
34	81.04	8.42	8.29	93.60	88.51	89.68	95.18	93.76	5.75
35	73.72	8.11	15.40	88.62	78.32	83.17	92.48	93.89	4.40
36	69.73	7.90	15.06	87.90	81.65	78.10	93.91	93.97	7.45
37	57.00	7.24	26.40	81.28	68.54	65.83	86.63	92.74	8.46
38	35.26	3.68	46.70	66.04	47.84	40.73	74.61	89.15	27.07
39	46.76	4.92	39.59	72.95	57.66	54.28	72.77	88.07	25.21
40	16.79	1.74	65.48	44.61	29.04	19.62	58.77	86.68	32.66
41	19.68	0.63	61.42	56.13	39.15	41.59	49.67	84.36	52.12
42	39.19	5.04	41.62	68.38	50.85	42.41	79.79	91.23	23.35
43	67.13	7.61	19.12	86.53	75.63	68.60	93.51	95.01	11.84
44	49.85	6.26	33.84	75.31	64.34	59.44	79.28	92.23	11.84
45	37.38	4.2	43.49	68.66	54.30	44.51	75.36	93.11	27.58
46	79.11	8.25	12.01	91.48	85.72	81.66	94.84	95.36	6.09
47	40.73	4.93	41.96	68.61	54.34	42.96	83.09	89.97	21.15
48	29.03	2.65	51.27	61.24	47.21	39.53	67.21	86.07	31.98
49	68.75	7.78	18.78	86.24	77.18	72.40	90.65	94.95	5.41
50	75.24	7.97	16.07	88.88	81.51	81.27	90.88	95.75	7.45
51	78.71	8.24	11.51	91.33	83.99	86.08	94.77	98.69	4.23
52	37.60	4.31	44.84	66.32	51.59	42.22	78.73	89.37	-6.6
53	73.20	8.07	14.38	88.78	79.55	77.39	93.93	93.73	2.2
54	67.43	7.67	20.30	85.64	75.90	70.28	87.57	94.91	12.07
55	70.06	7.90	18.44	87.29	76.67	78.93	90.79	93.49	7.11
56	71.88	7.83	17.77	88.24	78.07	77.51	89.21	97.74	8.46
57	80.05	8.3	11.00	91.81	85.72	83.94	95.03	96.12	2.88

In the 4th model, EBLEU, BLEU and NIST are significant and the level of significance for EBLEU and BLEU has increased, as well as their beta values, nonetheless the level of significance for NIST remains the same. However, at this stage of the model, METEOR becomes the most insignificant metric, therefore, it has been removed for the final (5th model). In comparison to the 2nd and 3rd models, in 4th model it has been observed that the BLEU is becoming more significant as the p-value ($p = 0.005$) is less than the alpha (0.05). Finally, in the last stage (5th model), the remaining metrics are

Table 84. Evaluation between ASR and original text.

SPKR	BLEU	NIST	TER	METEOR	METEOR-PL	EBLEU	RIBES	NER	RED
1	41.89	6.05	44.33	66.10	54.05	44.77	78.94	92.38	10.15
2	48.94	5.94	37.39	71.14	60.24	49.79	81.29	94.86	17.77
3	57.38	7.11	27.24	78.41	67.08	62.87	89.42	94.71	12.01
4	59.15	7.07	27.24	77.21	67.94	65.31	87.71	93.1	3.72
5	26.08	4.57	55.33	52.33	39.14	26.89	69.22	91.41	17.03
6	44.17	6.32	36.38	69.16	60.15	47.97	86.04	92.33	4.89
7	51.79	6.39	34.86	71.42	65.47	52.31	79.19	95.3	10.29
8	22.03	3.17	61.93	45.27	33.90	22.14	59.93	93.95	26.81
9	52.35	6.09	39.93	68.02	63.07	53.87	78.53	94.73	4.05
10	54.44	6.65	33.50	73.16	65.42	57.28	82.11	95.23	6.41
11	65.95	7.57	19.63	84.68	76.30	72.76	92.45	97.01	3.89
12	59.12	7.26	24.53	81.63	69.57	61.59	89.13	95.83	8.63
13	17.08	2.96	68.19	42.14	30.55	21.97	59.69	85.76	28.09
14	49.78	6.53	32.32	72.88	64.10	51.98	86.56	93.98	8.63
15	46.01	6.3	34.69	71.10	61.70	46.11	87.96	95.79	11.68
16	35.50	5.03	44.33	65.64	50.58	36.53	79.16	93.61	22.77
17	34.42	4.51	56.01	52.80	41.15	34.17	63.30	94.09	14.33
18	58.58	6.95	28.93	77.96	69.47	59.22	85.73	94.78	9.31
19	49.06	6.50	31.64	72.94	63.35	47.49	85.64	94.82	6.09
20	19.86	2.58	65.48	46.48	31.29	21.38	60.96	85.26	32.83

significant, as all of them have p-values less than alpha. Therefore, no next step of models has been executed. Here, the first model explains 75.9% of the variance for NER, however, the second model explains 76.3% of the variance of the NER, the 3rd model explains 76.6% of the variance of the NER, the 4th model explains 76.0% of the variance of the NER and the final model explains 76.1% of the variance of the NER. All these are statistically accepted Rsquare values.

From the regression, it has been found that BLEU is the most significant predictor of NER, after BLEU, NIST is significant and, finally, EBLEU is also a significant metric that can predict the NER better, thus, these can be alternatives to the NER metric. The regression equation that can compute the value of NER based on these three statistical metrics is:

$$NER = 86.55 + 0.254 * BLEU + 0.924 * NIST - 0.221 * EBLEU$$

Table 85. Regression result summary for NER and the six metrics.

Model		Unstandardized Coefficients		Standardized Coefficients	t	Sig.	Adjusted R-square
		B	Std. Error	Beta			
1st model	(Constant)	100.068	19.933		5.020	.000	
	BLEU	.176	.122	1.060	1.439	.157	
	NIST	1.162	.530	.738	2.192	.033	0.759
	TER	−.072	.186	−.353	−.386	.701	
	METEOR	−.205	.168	−.797	−1.216	.230	
	EBLEU	−.218	.078	−1.293	−2.806	.007	
	RIBES	−.067	.093	−.219	−.723	.473	
	METEOR_pl	.194	.160	.954	1.216	.230	
2nd model	(Constant)	92.622	4.895		18.920	.000	
	BLEU	.189	.117	1.136	1.614	.113	
	NIST	1.220	.504	.775	2.420	.019	0.763
	METEOR	−.190	.163	−.741	−1.170	.248	
	EBLEU	−.215	.077	−1.270	−2.803	.007	
	RIBES	−.051	.082	−.167	−.620	.538	
	METEOR_pl	.217	.147	1.067	1.479	.145	
3rd model	(Constant)	91.541	4.547		20.133	.000	
	BLEU	.184	.116	1.107	1.586	.119	
	NIST	1.068	.438	.678	2.440	.018	0.766
	METEOR	−.237	.143	−.923	−1.655	.104	
	EBLEU	−.193	.068	−1.145	−2.841	.006	
	METEOR_pl	.222	.146	1.093	1.527	.133	
4th model	(Constant)	90.471	4.550		19.885	.000	
	BLEU	.285	.096	1.716	2.955	.005	
	NIST	1.083	.443	.687	2.444	.018	0.760
	METEOR	−.099	.112	−.385	−.878	.384	
	EBLEU	−.204	.069	−1.210	−2.982	.004	
5th model	(Constant)	86.556	.913		94.814	.000	
	BLEU	.254	.090	1.531	2.835	.006	0.761
	NIST	.924	.404	.587	2.289	.026	
	EBLEU	−.221	.066	−1.310	−3.370	.001	

a. Dependent Variable: NER

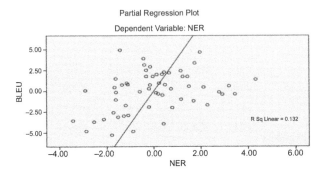

Figure 36. Partial regression plot for NER and BLEU.

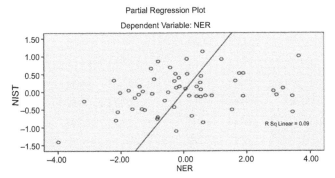

Figure 37. Partial regression plot for NER and NIST.

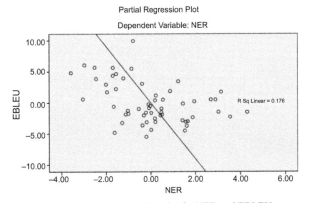

Figure 38. Partial regression plot for NER and EBLEU.

Moving forward, regression plots from the above regression for the significant metrics are presented below. These plots show the relation between dependent metric and each significant metric. The closer the dots in the plot are to the regression line, the better the R-square value and the better the relationship.

It is worth noting that the METEOR metric should be fine-tuned in order to work properly. This would, however, require more data; more research into this is planned for the future. Finally, the results presented in this monograph are derived from a very small data sample and may not be very representative of the task in general. The process of acquiring more data is still ongoing, so these experiments are going to be repeated once more data becomes available.

5.4.2 MT evaluation metrics within FACIT translation methodology

Differences in language and culture create the need for translators to convert text from one language into another. In order to preserve meaning, context must be analyzed in detail. Variation in the choice of words across languages is unavoidable but should be limited to the domain in question. This task is not trivial because translators have their own experiences that inform their personal translation preferences. This makes it necessary to develop accurate evaluation metrics that are independent of human experience, especially for the translation reconciliation step in PROMIS, and can provide expert reviews as additional information following backward translation. The development of such metrics and an analysis of their correlation with human judgment is the objective of this research.

In this research, we develop a semi-automatic semantic evaluation metric for Polish, based on the concept of the human-aided translation evaluation metric (HMEANT). The metric analyzes words and compares them with them others words; it takes into account synonyms, grammatical structure, syntax and grammar. Semantic analysis is conducted based on this information. The family of HMEANT metrics [256] has proven to be highly correlated with human judgment and accurate in the field of machine translation. We conducted an evaluation of the proposed metrics using a statistics-based support vector machine (SVM) [257] classifier. We trained it on Polish-to-English lecture translations on different topics taken from the Technology, Entertainment and Design (TED) Talks [15]. The quality of the translation was highly dependent on that of the input data and its similarity with the domain of the given topic. We carried out domain adaptation on the data using the Modified Moore Levis filtering and linear interpolation. We also applied deep neural networks in order to replicate the functions of the human brain. The literature has shown that neural networks can satisfactorily capture the semantics of written language. We trained a neural network on TED Talks' [15] data that had been previously adapted in order to learn translations that had been assessed as correct. In other words, our neural model attempts to capture different textual notions of cross-lingual similarity using input and output gates trying at the same time gates that belong to long and short-term memory (LSTM) architecture[44] [258] in order to build a Tree LSTM [259]. The network learning rate was set to 0.05 with a regularization strength of 0.0001. The number of dimensions of memory was set to 150 cells and training was performed for 10 epochs. We used well-known machine translation evaluation metrics BLEU [27], NIST [93], TER [188] and METEOR [31], and compared their results with one another as well as those of human judgment. In order to conduct such evaluation, it was necessary to train a statistical machine translation SMT [85] system and decode all original translations. For this purpose, we used the same data as for the SVM classifier [257]. All metrics were normalized in order to return values between 0 and 100, where 100 was considered to represent a perfect translation.

Our initial evaluative experiments were conducted on 150 official PROMIS[45] [260] questions and answers concerning physical capabilities, sleep disturbance and sleep-related impairment. We compared each candidate translation with our metrics for the original English questions, and checked to see if the Polish translation that had obtained a high score on our metrics had been chosen by the human reviewers. Based on this information, statistical analysis was conducted in order to count the correlations. We also analyzed the usefulness of the machine-based metrics. Our study provided adequate results to verify our initial assumption of the semantic HMEANT [256] metric, which is not influenced by human experience, habits and judgments. Nonetheless, samples on which experts disagreed were evaluated by other reviewers, and the subsequent results led us to conclude that the disagreements had been controversial. The neural network and the statistical classifier provided average results, much worse than those of HMEANT. The main problem was that the metrics were depended on bi-lingual data [261] required for initial training. PROMIS [260] translations are very specific, and finding high-quality, in-domain data in satisfactory quantities will likely improve performance.

5.4.2.1 PROMIS evaluation process

The entire PROMIS translation process was conducted using the FACIT methodology [262]. The methodology is illustrated briefly in Figure 39. It ensures that cross-lingual translations are highly consistent. The translation

---

[44] LSTM is a sequence learning technique that uses memory cells to preserve states over a long period of time, which allows distributed representations of sentences through distributed representations of words

[45] www.nihpromis.org

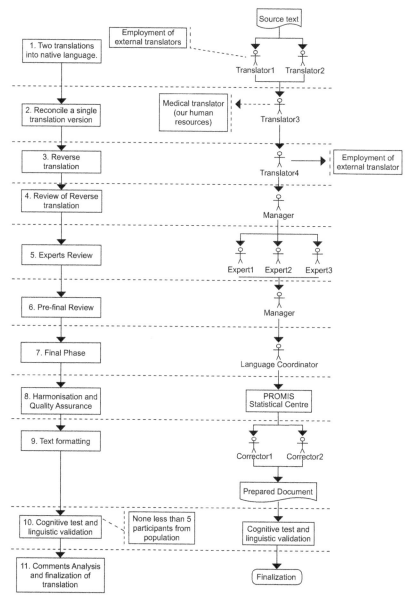

Figure 39. PROMIS translation methodology.

process can be divided into 11 steps. First, items in English are treated as source texts and translated simultaneously by two independent translators. It must be noted that the translators are native target language speakers. In the second step, a third independent translator (target language speaker as well) tries to generate a single translation based on the translations in the first step. This involves unifying translations into a hybrid version; in rare cases, the translator can recommend his/her own version. This translator also needs to outline reasons for the judgment and explain why the chosen translation conveys the correct meaning in the best possible way. Third, the reconciled version is backwardly translated into English by a native English translator. This person does not have knowledge of the original source items in English. In the fourth step the backward translation is reviewed. In this process, any possible discrepancy is identified and the intent of the translation is clarified. This is the initial step of the harmonization of languages. In the fifth step, three independent experts (native speakers of the target language) examine all previously taken actions in order to select the most appropriate translation for each item or provide their own proposals. The experts must be either linguists or healthcare professionals (preferably a mixed group). In the sixth step, the translation project manager evaluates the merit of expert comments and, if necessary, identifies problems. Using this information, the manager formulates guidance for the target language coordinator. This coordinator in the seventh step determines the final translation by examining all information gathered in the previous steps. He/She is required to explain and justify the decisions, especially if the final version is different from the reconciled one. In order to ensure cross-language harmonization and quality assurance, the translation project manager in the eighth step performs a preliminary assessment of the accuracy and equivalence of the final translation with respect to the original. A quality review is also conducted by the PROMIS Statistical Center in order to verify consistency with other languages, if necessary. In the ninth step, the items are formatted and proofread independently by two native speakers and the results are reconciled. The target language version is pre-tested within the group of target language native speakers in the tenth step. Each is debriefed and interviewed in order to check if his/her intended meaning was correctly understood. In the final step, comments following pre-testing are analyzed and any possible issues summarized. If necessary, new solutions are proposed [263].

5.4.2.2 *Preparation of automatic evaluation metrics and data*

In this section, we briefly characterize the evaluation metrics developed, together with how the machine translation system was prepared. We also describe metrics that are already commonly used in machine translation—BLEU, NIST, TER and METEOR. For these four metrics, together with the SVM classifier and neural network preparation, we used data from the TED talks as a parallel corpus (approximately 15 MB) which contained almost 2 million words that had not been tokenized. The talks' transcripts were provided as pure text encoded in UTF-8, and the transcripts were prepared by the FBK team. They also separated the transcripts into sentences (one per line) and aligned the language pairs. It should be emphasized that both the automatic and manual pre-processing of these training data were required. The extraction of transcription data from the XML files ensured an equal number of lines for English and Polish. However, some discrepancies in text parallelism could not be avoided. These discrepancies were mainly repetitions of the Polish text not included in the English text. Another problem was that TED data contained many errors. We first considered spelling errors that artificially increased dictionary size and worsened the statistics of the translation phrase table. This problem was solved using the tool proposed in [18] and manual proofreading. The final TED corpus consisted of 92,135 unique Polish words.

The quality of machine translation is highly dependent on that of the input data and their similarity with the domain of the given topic. We conducted domain adaptation on the TED data using the Modified Moore Levis filtering and linear interpolation in order to adapt it for the assessment of translation in PROMIS [264].

The problem of data sparsity, if not addressed, can lead to low translation accuracy, false errors in evaluation and low classifier scores. The quality of domain adaptation depends heavily on the training data used to optimize the language and translation models in an SMT system. The selection and extraction of domain-specific training data from a large, general corpus addresses this issue [113]. This process uses a parallel, general domain corpus and a general domain monolingual corpus in the target language. The result is a pseudo-in-domain sub-corpus. As described by Wang et al. [265], there are in general three processing stages in data selection for domain adaptation. First, sentence pairs from the parallel, general domain corpus are scored for relevance to the target domain. Second, resampling is performed to select the best-scoring sentence pairs to retain in the pseudo-in-domain sub-corpus. These two steps can also be applied to the general-domain monolingual corpus in order to select sentences for use in a language model. Third, after collecting a substantial number of sentence pairs (for the translation model) or sentences (for the language model), the models are trained on the sub-corpus that represents the target domain [265]. Similarity measurement is required in order to select sentences for the pseudo-in-domain sub-corpus. Three state-of-the art approaches were used for similarity measurement. The cosine tf-idf criterion looks for word overlap in order to determine similarity. This technique is specifically helpful in reducing the number of out-of-vocabulary (OOV) words, but is sensitive to noise in the data. A perplexity-based criterion considers the n-gram word order in addition to collocation. Lastly, edit distance simultaneously considers word order, position and overlap. It is the strictest of the three approaches. In their study [265], Wang et al. found that a combination of these approaches provided the best performance in domain adaptation for Chinese-to-English corpora [265]. In accordance with Wang et al.'s approach [265], we use a combination of the criteria in both the corpora and the language models followed by Modified Moore Levis filtering and linear interpolation for the language models. The three similarity metrics are used to select different pseudo-in-domain sub-corpora. The sub-corpora are then joined during resampling, based on a combination of the three metrics. Similarly, the three metrics are combined for domain adaptation during translation. We empirically found acceptance rates that allowed us to harvest only 20% of most domain-similar data [265]. As in-domain data, we used the original PROMIS translations.

5.4.2.3 *SMT system preparation*

The machine translation system is needed to automatically translate English source documents and compare the results of the translation with human judgments using the widely used BLEU, NIST, TER and METEOR metrics. These metrics are the only possible means of extracting cross-language information.

Our statistical machine translation (SMT) system was implemented using the Moses open-source SMT toolkit with its Experiment Management System (EMS) [113]. The phrase symmetrization method is set to grow-diag-final-and for word alignment processing. Two-way direction alignments are first intersected, such that only alignment points occurring in both alignments remain. In the second phase, additional alignment points in their union are added. A growing step adds potential alignment points of unaligned words and neighbors. The neighborhood can be set directly to left, right, top or bottom, as well as diagonal (grow-diag). In the final step, alignment points between words from which at least one is unaligned are added (grow-diag-final). If the grow-diag-final-and method is used, an alignment point between two unaligned words appears [19]. SyMGiza++, a tool that supports the creation of symmetric word alignment models, is used to extract parallel phrases from the data. This tool enables alignment models that support many-to-one and one-to-many alignments in both directions between language pairs. SyMGiza++ is also designed to leverage the power of multiple processors through advanced threading management, thus making it very fast. Its alignment process uses four models during training in order to progressively refine the alignment results. This approach yielded impressive results in Junczys-Dowmunt and Szał [266]. Out-of-vocabulary (OOV) words pose another significant challenge to SMT systems. If not addressed, unknown words appear, untranslated, in the output, thus lowering translation quality. To address OOV words, we implemented the unsupervised transliteration model (UTM) in the Moses toolkit. The UTM is an unsupervised, language-independent approach for learning OOV words. We used the post-decoding transliteration option in this tool. UTM uses a transliteration phrase translation table to evaluate and score multiple possible transliterations [267, 80]. The KenLM tool is also applied to the language model to train in addition to binarization. This library enables highly efficient queries on the language models, saving memory and computation time. The lexical values of phrases are used to condition their reordering probabilities. We used KenLM with lexical reordering set to hier-msd-bidirectional-fe. This setting uses a hierarchical model that considers three orientation types based on both source and target phrases: Monotone (M), swap (S) and discontinuous (D). The probabilities of the possible phrase orders are examined by a bidirectional reordering model [10, 47, 268].

5.4.2.4 *Support vector machine classifier evaluation*

In machine learning, support vector machines are supervised learning models with associated learning algorithms that analyze data used for classification and regression analysis. Given a set of training examples, each marked as belonging to one of two categories, an SVM training algorithm builds a model that assigns new examples to one category or the other, hence making it a non-probabilistic linear binary classifier. An SVM model is a representation of examples as points in space, mapped so that instances of the categories are divided by a clear gap that is as wide as possible. New examples are then mapped into that space and predicted to belong to one of the categories based on the side of the gap in which they reside.

For the sentence similarity metric, we implemented an algorithm that normalizes the likelihood output of a statistical SVM classifier into the range 0–1. The classifier must be trained to determine if sentence pairs are translations of each other and evaluate them. In addition to being an excellent classifier, the SVM can provide a distance in the separation hyperplane during classification that can be easily modified using a sigmoid function to return a value similar to the likelihood between 0 and 1 [24]. The use of the classifier means that the quality of the alignment depends not only on the input, but also on the quality of the trained classifier. To train the classifier, good quality parallel data are needed as well as a dictionary that included translation probabilities. For this purpose, we used the TED talks [3] corpora. To obtain a dictionary, we trained a phrase table and extracted one-grams from it [113].

The quality of evaluation is defined by a trade-off between precision and recall. The classifier has two configurable variables [269]:

- Threshold: The confidence threshold to accept a translation as "good." A lower value means higher precision and lower recall. "Confidence" is a probability estimated from an SVM classifying "is a good translation" or "is a poor translation".[46]

- Penalty: Controls the magnitude of "skipping ahead" allowed in translation quality [270]. Say we are evaluating subtitles, where there are few or no extra paragraphs, and the alignment should be more or less one-to-one; then, the penalty should be high. If aligning entities that are moderately good translations of one another, where there are extra paragraphs for each language, the penalty should be lower.

Both these parameters are selected automatically during training, but can be manually adjusted if necessary.

5.4.2.5 *Neural network-based evaluation*

Neural machine translation and evaluation are new approaches to machine translation where a large neural network is trained to maximize translation "understanding" performance. This is a radical departure from existing phrase-based statistical translation approaches, where a translation system consists of subcomponents that are separately optimized. A bidirectional recurrent neural network (RNN), known as an encoder, is used by the neural network to encode a source sentence for a second RNN, known as the decoder, which is used to predict words in the target language [240].

Having created it, we start training the network. Both human and neural networks can learn and develop to be experts in a certain domain, but the major difference is that humans can forget, unlike neural networks. A neural network never forgets, once trained, as information learned is permanent and hard coded; human knowledge may not be permanent, and is influenced by personal experience. Another major difference is accuracy. Once a particular process is automated via a neural network, the results can be repeated, and will remain as accurate as in the first calculation. The first 10 processes may be accurate, with the next 10 characterized by mistakes due to, for instance, personal conditions. Another major difference is speed [241].

Our neural network was implemented using the Groundhog and Theano tools. Most neural machine translation models that have been proposed belong to the encoder-decoder family [243], with an encoder and a decoder for every language, or a language-specific encoder for each sentence application whose

---

[46] https://github.com/machinalis/yalign/issues/3

outputs are compared [243]. A translation is the output a decoder receives from the encoded vector. The entire encoder-decoder system is jointly trained to maximize the correct evaluations. A significant and unique feature of this approach is that it does not attempt to encode an input sentence into a vector of fixed length. Instead, the sentence is mapped to a vector sequence, and the model adaptively chooses a vector subset as it decodes the translation. To be more precise, we trained a neural network on TED Talks' data that had been previously adapted to teach it correctly evaluated translations. Our neural model tries to capture different textual notions of cross-language similarity using input and output gates in the same time gates as the LSTM [258] architecture to build Tree LSTMs. The network learning rate was set to 0.05 with a regularization strength of 0.0001. The number of dimensions of the memory was set to 150 cells, and training was performed for 10 epochs.

5.4.2.6 Results

In order to verify the predictive power of each metric, we evaluated (using the official English version) each using two candidate translations and a third reconciled version. We also analyzed the coverage of the results of the metrics given the final translation following the entire PROMIS evaluation process. The assessment was conducted on 150 official PROMIS questions concerning physical capabilities, sleep disturbance and sleep-related impairments. The weighted index method [271] was used to rank the different pools of the translation process in comparison with human judgment. Pools were compared with human judgment in terms of how the translations were judged by humans and assigned a weight for each case. The weighted index value based on the relative weights of the chosen scale was computed using the following formula:

$$PI_x = \sum (Wi\ fi)/N \ \dots\dots\ (\textbf{Formula 1})$$

where,

PI_x = pool index value

fi = frequency of cases of match or mismatch

Wi = weight of the rating from lowest to highest

N = summation of the frequency of cases (N = 150)

An appropriate weight was assigned to the different attributes given by the translation; each pool had three columns, and was matched with the three columns representing human judgment. The weight was assigned based on the level of matching, and is summarized in Table 86. Note that double matches occurred when the reconciled version was identical to the proposed one.

Table 86. Weights of categories.

Category	Weight
Double match with human judgment	4
One match with human judgment	3
One mismatch with human judgment	3
Double mismatch with human judgment	4

Based on the level of match, the frequency of each pool as matched with human judgment was found for double matches, single matches or mismatches with human judgment. The frequency of each translation pool is presented in Table 87.

Table 87. Frequency of matches with human judgment for different translations.

Level of match	HMEANT	BLEU	NIST	METEOR	TER	NEURAL	SVM
Double mismatch with human judgment	3	4	11	8	7	3	5
One mismatch with human judgment	27	85	73	81	86	53	70
One match with human judgment	112	54	66	58	53	86	69
Double match with human judgment	8	7	0	3	4	8	6

Using the frequency and the formula, the weighted index for each pool was estimated; for example, for BLEU, the weighted index was:

$$PI_{BLEU} = [(3 \times 1) + (27 \times 2) + (112 \times 3) + (8 \times 4)]/150 = 2.42$$

Using the same process, the index values for each pool were calculated and are shown in Figure 40.

We know that HMEANT best matches human judgment and, therefore, was the best evaluation pool. However, we explored the translation pools that yielded the best prediction of HMEANT. In order to find this relation, backward stepwise linear regression [272] was applied to matched values of each pool. The results are presented in Table 88.

As shown in Table 88, in the first model, NIST was the most insignificant metric in predicting HMEANT, and was, therefore, removed from the second model. In the second model, SVM was the most insignificant and was removed in the next model. In the third model, neural was the most insignificant matrix; finally, in the fourth model, METEOR was found to be insignificant for the prediction of HMEANT; therefore, in the fifth model, only BLEU and TER

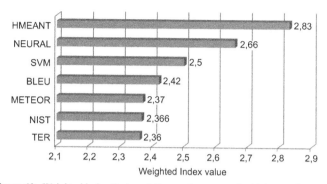

Figure 40. Weighted index for translation pools compared with human judgment.

were retained, and had p-values of less than 0.05, indicating that they were significant for the prediction of HMEANT.

According to the index calculation, the closer the value to 4 (the highest weight), the better the match and, therefore, the better the translation quality. As shown in Figure 40, the highest index value was observed for HMEANT, which confirmed that it is the best of all translation pools and closest to human judgment. After HMEANT, neural had the second-highest score, whereas SVM had the third-highest. This implied that after HMEANT, neural best matched human judgment (the second pool). In addition, SVM was the third-best match with human judgment. In this case, BLEU and METEOR had moderate levels of matches, and NIST and TER had the lowest match with human judgment.

From the stepwise linear regression model, we found that both BLEU and TER were significant predictors of HMEANT. Between BLEU and TER, the former had a stronger positive relation with HMEANT with a beta value ($B_{BLEU} = 0.209$). TER also showed a strong relation with the beta value ($B_{TER} = -0.385$). This shows that with increasing TER values, the HMEANT values decreased. In this case, the values were computed based on the matching score, values in relation to human judgment. The regression equation to compute the value of HMEANT based on TER and BLEU statistical metrics is:

$$HMEANT = 3.235 + 0.209 * BLEU - 0.385 * TER \dots \dots . \text{(Equation 1)}$$

In summation, we proved that automatic and semi-automatic evaluation metrics can properly predict human judgments. HMEANT is a very good predictor of human judgments. The slight difference between the metric and human judgements were obtained because of human translation habits and knowledge-dependent decisions, since the choices made by translators were controversial. Even though the SVM and the neural network yielded much

Table 88. HMEANT results.

Model B		Unstandardized Coefficients		Standardized Coefficients	t	Sig.	Adjusted R-square
		Std. Error	Beta				
1	(Constant)	3.395	.388		8.750	.000	0.23
	BLEU	.201	.063	.235	3.204	.002	
	NIST	−.004	.063	−.005	−.068	.946	
	METEOR	−.087	.065	−.101	−1.340	.182	
	TER	−.405	.065	−.466	−6.257	.000	
	NEURAL	.026	.064	.030	.408	.684	
	SVM	.023	.063	.027	.358	.721	
2	(Constant)	3.382	.340		9.938	.000	0.244
	BLEU	.201	.062	.235	3.243	.001	
	METEOR	−.087	.065	−.100	−1.345	.181	
	TER	−.405	.065	−.466	−6.282	.000	
	NEURAL	.026	.064	.030	.412	.681	
	SVM	.022	.063	.026	.356	.723	
3	(Constant)	3.430	.312		10.983	.000	0.248
	BLEU	.198	.061	.232	3.233	.002	
	METEOR	−.083	.064	−.096	−1.311	.192	
	TER	−.400	.063	−.460	−6.366	.000	
	NEURAL	.025	.063	.028	.389	.698	
4	(Constant)	3.478	.286		12.177	.000	0.253
	BLEU	.201	.061	.235	3.298	.001	
	METEOR	−.080	.063	−.092	−1.269	.206	
	TER	−.400	.063	−.459	−6.374	.000	
5	(Constant)	3.235	.212		15.231	.000	0.250
	BLEU	.209	.061	.245	3.447	.001	
	TER	−.385	.062	−.443	−6.235	.000	

[a] Dependent Variable: HMEANT

lower accuracy, it should be noted that both methods are dependent on the reference training corpus, which was not well-suited for such an evaluation (but was the best we could obtain). Training such metrics on larger amounts of in-domain data can significantly alter the situation.

Unfortunately, predicting HMEANT is only semi-automatic, and is not time or cost effective. While it provides reliable and human habit-independent results, it still requires a lot of annotation work. This is why the prediction of this metric, along with the BLEU- and TER-dependent equation, was an interesting finding. These automatic metrics can provide robust and accurate results or, at least, initial results that can assist human reviewers.

5.4.3 Augmenting SMT with semantically-generated virtual-parallel data

Statistical machine translation (SMT) is a methodology based on statistical data analysis. The performance quality of SMT systems largely depends on the quantity and quality of the parallel data used by these systems; that is, if the quantity and quality of the parallel data are high, this will boost the SMT results. Even so, parallel corpora remain scarce and are not easily available [273]. Moreover, in order to increase SMT performance, the genre and language coverage of the data should be limited to a specific text domain, law or medical texts. In particular, little research has been conducted on languages with few native speakers and, thus, with a limited audience, even though most existing human languages are spoken by only a small population of native speakers, as shown in Table 89.

Despite the enormous number of people with technological knowledge and access, many are excluded because they cannot communicate globally due to language divides. Consistent with Anderson et al. [274], over 6,000 languages [274] are used globally; there is no universal spoken language for communication. The English language is only the third most popular (used by only 5.52% of the global population); Spanish (5.85%) and Mandarin (14.1%) are more common [275]. Moreover, fewer than 40% of citizens of the European Union (not including developing or Eastern European countries) know English [276], which makes communication a problem even within the EU [277].

This has created a technical gap between languages that are widely spoken in comparison to languages with few speakers. This also led to a big gap between the quality and amount of available parallel corpora for less common language pairs, which makes natural language processing sciences slower in such countries.

As a result, high-quality data exist for just a few language pairs in particular domains (e.g., Czech-English law texts domain), whereas the majority of languages lack sufficient linguistic resources, such as parallel data for good quality research or natural language processing tasks. Building a translation system that can handle all possible language translations would require millions of translation directions and a huge volume of parallel data. Moreover, if we consider multiple domains in the equation, the requirements for corpus training in machine translation increase dramatically. Thus, the current study explored methods to build a corpus of high-quality parallel data, using Czech-English as the language pair.

Table 89. Top languages by population: Asterisks mark the 2010 estimates for the top dozen languages.

Rank	Language	Native speakers in millions 2007 (2010)	Fraction of world population (2007)	Rank	Language	Native speakers in millions 2007 (2010)	Fraction of world population (2007)
1	Mandarin (entire branch)	935 (955)	14.1%	51	Igbo	24	0.36%
2	Spanish	390 (405)	5.85%	52	Azerbaijani	23	0.34%
3	English	365 (360)	5.52%	53	Awadhi	22[4]	0.33%
4	Hindi [Note 1]	295 (310)	4.46%	54	Gan Chinese	22	0.33%
5	Arabic	280 (295)	4.23%	55	Cebuano (Visayan)	21	0.32%
6	Portuguese	205 (215)	3.08%	56	Dutch	21	0.32%
7	Bengali (Bangla)	200 (205)	3.05%	57	Kurdish	21	0.31%
8	Russian	160 (155)	2.42%	58	Serbo-Croatian	19	0.28%
9	Japanese	125 (125)	1.92%	59	Malagasy	18	0.28%
10	Punjabi	95 (100)	1.44%	60	Saraiki	17[5]	0.26%
11	German	92 (95)	1.39%	61	Nepali	17	0.25%
12	Javanese	82	1.25%	62	Sinhalese	16	0.25%
13	Wu (inc. Shanghainese)	80	1.20%	63	Chittagonian	16	0.24%
14	Malay (inc. Malaysian and Indonesian)	77	1.16%	64	Zhuang	16	0.24%
15	Telugu	76	1.15%	65	Khmer	16	0.24%
16	Vietnamese	76	1.14%	66	Turkmen	16	0.24%
17	Korean	76	1.14%	67	Assamese	15	0.23%
18	French	75	1.12%	68	Madurese	15	0.23%
19	Marathi	73	1.10%	69	Somali	15	0.22%
20	Tamil	70	1.06%	70	Marwari	14[4]	0.21%
21	Urdu	66	0.99%	71	Magahi	14[4]	0.21%
22	Turkish	63	0.95%	72	Haryanvi	14[4]	0.21%
23	Italian	59	0.90%	73	Hungarian	13	0.19%
24	Yue (incl. Cantonese)	59	0.89%	74	Chhattisgarhi	12[4]	0.19%
25	Thai (excl. Lao)	56	0.85%	75	Greek	12	0.18%

Table 89 cont. ...

...*Table 89 cont.*

Rank	Language	Native speakers in millions 2007 (2010)	Fraction of world population (2007)	Rank	Language	Native speakers in millions 2007 (2010)	Fraction of world population (2007)
26	Gujarati	49	0.74%	76	Chewa	12	0.17%
27	Jin	48	0.72%	77	Deccan	11	0.17%
28	Southern Min (incl. Fujianese/ Hokkien)	47	0.71%	78	Akan	11	0.17%
29	Persian	45	0.68%	79	Kazakh	11	0.17%
30	Polish	40	0.61%	80	Northern Min	10.9	0.16%
31	Pashto	39	0.58%	81	Sylheti	10.7	0.16%
32	Kannada	38	0.58%	82	Zulu	10.4	0.16%
33	Xiang (Hunnanese)	38	0.58%	83	Czech	10.0	0.15%
34	Malayalam	38	0.57%	84	Kinyarwanda	9.8	0.15%
35	Sundanese	38	0.57%	85	Dhundhari	9.6[4]	0.15%
36	Hausa	34	0.52%	86	Haitian Creole	9.6	0.15%
37	Odia (Oriya)	33	0.50%	87	Eastern Min	9.5	0.14%
38	Burmese	33	0.50%	88	Ilocano	9.1	0.14%
39	Hakka	31	0.46%	89	Quechua	8.9	0.13%
40	Ukrainian	30	0.46%	90	Kirundi	8.8	0.13%
41	Bhojpuri	29[4]	0.43%	91	Swedish	8.7	0.13%
42	Tagalog/ Filipino	28	0.42%	92	Hmong	8.4	0.13%
43	Yoruba	28	0.42%	93	Shona	8.3	0.13%
44	Maithili	27[4]	0.41%	94	Uyghur	8.2	0.12%
45	Uzbek	26	0.39%	95	Hiligaynon/ Ilonggo (Visayan)	8.2	0.12%
46	Sindhi	26	0.39%	96	Mossi	7.6	0.11%
47	Amharic	25	0.37%	97	Xhosa	7.6	0.11%
48	Fula	24	0.37%	98	Belarusian	7.6[6]	0.11%
49	Romanian	24	0.37%	99	Balochi	7.6	0.11%
50	Oromo	24	0.36%	100	Konkani	7.4	0.11%
				Total		**5,610**	**85%**

Multiple studies have been performed in order to automatically acquire additional data for enhancing SMT systems in the long term [278, 7]. All such approaches have focused on discovering authentic text from real-world sources for both the source and target languages. However, our study presents an alternative approach for building this parallel data. In creating virtual parallel data, as we might call it, at least one side of the parallel data is generated, for which purpose we use monolingual text (news internet crawl in Czech, in this case). For the other side of the parallel data, we use an automated procedure to obtain a translation of the text. In other words, our approach generates rather than gathers parallel data. To monitor the performance and quality of the automatically generated parallel data and to maximize its utility for SMT, we focus on compatibility between the diverse layers of an SMT system.

It is recommended that an estimate be considered reliable when multiple systems show a consensus on it. However, since the output of machine translation (MT) is human language, it is much too complicated to seek unanimity from multiple systems in order to generate the same output each time we execute the translation process. In such situations, we can choose partial compatibility as an objective rather than complete agreement between multiple systems. To evaluate the generated data, we can use the Levenshtein distance as well as implementing a back-translation procedure. Using this approach, only those pairs that pass an initial compatibility check, when translated back into the native language and compared to the original sentences, will be accepted.

We can use this method to easily generate additional parallel data from monolingual news data provided for WMT16.[47] Retraining the newly assessed data during this procedure enhances translation system performance. Moreover, linguistic resource pairs that are rare can be improved. This methodology is not limited to languages but is also very significant for rare but important language pairs. Most significantly, the virtual parallel corpus generated by the system is applicable to MT as well as other natural language processing (NLP) tasks.

5.4.3.1 Previous research

In this study, we present an approach based on generating comprehensive multilingual resources through SMT systems. We are now working on two approaches for MT applications: Self-training and translation via bridge languages (also called "pivot languages"). These approaches are different from those discussed previously: While self-training is focused on exploiting

---

[47] http://www.statmt.org/wmt16/translation-task.html

the available bilingual data, to which the linguistic resources of a third language are rarely applied, translation via bridge languages focuses more on correcting the alignment of the prevailing word segment. This latter approach also incorporates the phrase model concept rather than exploring the new text in context, by examining translations at the word, phrase, or even sentence level, through bridge languages. The methodology of this monograph lies in between the paradigm of self-training and translating via a bridge language. Our study generates data instead of gathering information for parallel data, while we also apply linguistic information and inter-language relationships to eventually produce translations between the source and target languages.

Callison-Burch and Osborne [279] presented a cooperative training method for SMT that comprises the consensus of several translation systems to identify the best translation resource for training. Similarly, Ueffing et al. [280] explored model adaptation methods to use monolingual data from a source language. Furthermore, as the learning progressed, the application of that learned material was constrained by a multi-linguistic approach without introducing new information from a third language.

In another approach, Mann and Yarowsky [281] presented a technique to develop a translation lexicon based on transduction models of cognate pairs through a bridge language. In this case, the edit distance rate was applied to the process rather than the general MT system of limiting the vocabulary range for majority European languages. Kumar et al. [282] described the process of boosting word alignment quality using multiple bridge languages. In Wu and Wang [283] and Habash and Hu [284], phrase translation tables were improved using phrase tables acquired in multiple ways from pivot languages. In Eisele et al. [285], a hybrid method was combined with RBMT (Rule-Based Machine Translation) and SMT systems. This methodology was introduced to fill gaps in the data for pivot translation. Cohn and Lapata [286] presented another methodology to generate more reliable results of translations by generating information from small sets of data using multi-parallel data.

Contrary to the existing approaches, in this study, we returned to the black-box translation system. This means that virtual data could be widely generated for translation systems, including rule-based, statistics-based and human-based translations. The approach introduced in Leusch et al. [287] pooled the results of translations of a test set created by any of the pivot MTs per unique language. However, this approach was not found to enhance the systems, therefore, the novel training data were not used. Amongst others, Bertoldi et al. [288] also conducted research on pivot languages, but did not consider applying universal corpus filtering, which is the measurement of compatibility to control data quality.

5.4.3.2 *Generating virtual parallel data*

To generate new data, we trained three SMT systems based on TED, QED and News Commentary corpora. The Experiment Management System [8] from the open source Moses SMT toolkit was utilized to carry out the experimentation. A 6-gram language model was trained using the SRI Language Modeling toolkit (SRILM) [9]. Word and phrase alignment was performed using the SyMGIZA++ symmetric word alignment tool [266] instead of GIZA++. Out-of-vocabulary (OOV) words were monitored using the Unsupervised Transliteration Model [79]. Working with the Czech (CS) and English (EN) language pair, the first SMT system was trained on TED [15], the second on the Qatar Computing Research Institute's Educational Domain Corpus (QED) [193], and the third using the News Commentary corpora provided for the WMT16[48] translation task. Official WMT16 test sets were used for system evaluation. Translation engine performance was measured by the BLEU metric [27]. The performance of the engines is shown in Table 90.

Table 90. Corpora used for generation of SMT systems.

Corpus	Direction	BLEU
TED	CS→EN	16.17
TED	EN→CS	10.11
QED	CS→EN	23.64
QED	EN→CS	21.43
News Commentary	CS→EN	14.47
News Commentary	EN→CS	9.87

All engines worked in accordance with Figure 41, and the Levenshtein distance was used to measure the compatibility between translation results. The Levenshtein distance measures the diversity between two strings. Moreover, it also indicates the edit distance and is closely linked to the paired arrangement of strings [289].

Mathematically, the Levenshtein distance between two strings a, b [of length |a| and |b|, respectively] is given by $lev_{a,b}$ [|a|, |b|] where:

$$
lev_{a,b}(i,j) = \begin{cases} \max(i,j) & \text{if } \min(i,j) = 0 \\ \min \begin{cases} lev_{a,b}(i-1,j) + 1 \\ lev_{a,b}(i,j-1) + 1 \\ lev_{a,b}(i-1,j-1) + 1_{[a_i \neq b_j]} \end{cases} & otherwise. \end{cases}
$$

---

[48] http://www.statmt.org/wmt16/

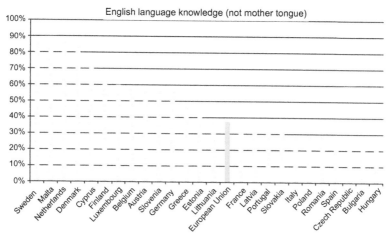

Figure 41. English language knowledge.

In this equation, $1_{[ai \neq bj]}$ is the display function, equal to 0 when $a_i = b_j$ and equal to 1 otherwise, and $lev_{a,b}[i, j]$ is the distance between the first i characters of a and the first j characters of b.

Using the combined methodology and monolingual data, parallel corpora were built. Statistical information on the data is provided in Table 91.

Table 91. Specification of generated corpora.

Data set	Number of Sentences		Number of Unique Czech Tokens	
	Monolingual	Generated	Monolingual	Generated
News 2007	100,766	83,440	200,830	42,954
News 2008	4,292,298	497,588	2,214,356	168,935
News 2009	4,432,383	527,865	2,172,580	232,846
News 2010	2,447,681	269,065	1,487,500	100,457
News 2011	8,746,448	895,247	2,871,190	298,476
News 2012	7,538,499	849,469	2,589,424	303,987
News 2013	8,886,151	993,576	2,768,010	354,278
News 2014	8,722,306	962,674	2,814,742	322,765
News 2015	8,234,140	830,987	2,624,473	300,456
TOTAL	53,366,020	5,944,583	19,743,105	2,125,154

The purpose of this research was to create synthetic parallel data to train a machine translation system by translating monolingual texts with multiple machine translation systems and various filtering steps. This objective is not new; synthetic data have been created in the past. However, the novel aspect of the present monograph is its use of three MT systems, application of the Levenshtein distance between their outputs as a filter, and—much more importantly—its use of back-translation as an additional filtering step. In Table 92, we show statistical information on the corpora used without the back-translation step.

Table 92. Specification of generated corpora without back-translation.

Data set	Number of Sentences		Number of Unique Czech Tokens	
	Monolingual	Generated	Monolingual	Generated
News 2007	100,766	93,342	200,830	120,654
News 2008	4,292,298	1,654,233	2,214,356	1,098,432
News 2009	4,432,383	1,423,634	2,172,580	1,197,765
News 2010	2,447,681	1,176,022	1,487,500	876,654
News 2011	8,746,448	2,576,253	2,871,190	1,378,456
News 2012	7,538,499	2,365,234	2,589,424	1,297,986
News 2013	8,886,151	2,375,857	2,768,010	1,124,278
News 2014	8,722,306	1,992,876	2,814,742	1,682,673
News 2015	8,234,140	2,234,987	2,624,473	1,676,343
TOTAL	53,366,020	15,892,438	19,743,105	10,453,241

5.4.3.3 Semantically-enhanced generated corpora

The artificially generated corpora presented in Table 92 were obtained using statistical translation models, which are based purely on how frequently "things" happen, and not on what they really mean. This means that they do not really understand what was translated. In this research, these data were additionally extended with semantic information so as to improve the quality and scope of the data domain. The word relationships were integrated into generated data using the WordNet database.

The way in which WordNet was used to obtain a probability estimator was shown in Cao et al. [290]. In particular, we wanted to obtain $P(w\_i|w)$, where w_i and w are assumed to have a relationship in WordNet. The formula is as follows:

$$P(w\_i|w) = \frac{c(w\_i,w|W,L)}{\sum_{w\_j} c(w\_j,w|W,L)}$$

where W is a window size and $c(w\_i,w|W,L)$ is the count of w_i and w appearing together within W-window. This can be obtained simply by counting each within a certain corpus. In order to smooth the model, we applied interpolated Kneser-Ney [32] smoothing strategies.

The following relationships were considered: Synonym, hypernym, hyponym and hierarchical distance between words.

In Table 93, we show statistical information on the semantically enhanced corpora produced previously and shown in Table 91.

Table 93. Specification of semantically generated corpora without back-translation.

Data set	Number of Sentences		Number of Unique Czech Tokens	
	Monolingual	Generated	Monolingual	Generated
News 2007	100,766	122,234	200,830	98,275
News 2008	4,292,298	1,467,243	2,214,356	803,852
News 2009	4,432,383	1,110,234	2,172,580	959,847
News 2010	2,447,681	982,747	1,487,500	585,852
News 2011	8,746,448	1,397,975	2,871,190	1,119,281
News 2012	7,538,499	1,759,285	2,589,424	968,975
News 2013	8,886,151	1,693,267	2,768,010	982,948
News 2014	8,722,306	1,462,827	2,814,742	1,243,286
News 2015	8,234,140	1,839,297	2,624,473	1,273,578
TOTAL	53,366,020	11,835,109	19,743,105	8,035,470

Another common approach to semantic analysis that is also used within this research is latent semantic analysis (LSA). LSA has already been shown to be very helpful in automatic speech recognition (ASR) [291] and many other applications, which was the reason for incorporating it within the scope of this research. The high-level idea of LSA is to convert words into concept representations and to assume that if the occurrence of word patterns in documents is similar, then the words are also similar. The mathematical model can be defined as follows:

In order to build the LSA model, a co-occurrence matrix W will first be built, where w_{ij} is a weighted count of word w_j and document d_j.

$$w_{ij} = G_i L_{ij} C_{ij}$$

where C_{ij} is the count of w_i in document d_j; L_{ij} is local weight; and G_i is global weight. Usually, L_{ij} and G_i can use TF/IDF.

Then, singular value decomposition (SVD) analysis will be applied to W, as

$$W = U S V^T$$

where W is a M*N matrix (M is vocabulary size, N is document size); U is M*R, S is R*R, and V is a R*N matrix. R is usually a predefined dimension number between 100 and 500.

After that, each word w_i can be denoted as a new vector $U_i = u_i*S$. Based on this new vector, the distance between two words is defined as:

$$K(U_i, U_j) = \frac{u_i*S^2*u_m^T}{|u_i*S|*|u_m*S|}$$

Therefore, clustering can be performed to organize words into K clusters, $C\_1, C\_2, ..., C\_K$.

If $H\_\{q-1\}$ is the history for word W_q, then it is possible to obtain the probability of W_q given $H\_\{q-1\}$ using the following formula:

$$P(W\_q|H\_\{q-1\}) = P(W\_q|W\_\{q-1\}, W\_\{q-2\}, ... W\_\{q-n+1\}, d\_\{q\_1\})$$

$$= P(W\_q|W\_\{q-1\}, W\_\{q-2\}, ... W\_\{q-n+1\}) * P(W\_q|d\_\{q\_1\}|)$$

where $P(W\_q|W\_\{q-1\}, W\_\{q-2\}, ... W\_\{q-n+1\})$ is the N-gram model; $P(d\_\{q\_1\}|W\_q)$ is the LSA model.

Additionally,

$$P(W\_q|d\_\{q\_1\}) = P(U\_q|V\_q) = K(U\_q, V\_\{q\_1\})/Z(U,V) K(U\_q, V\_\{q\_1\})$$
$$= \backslash frac\{U\_q * S * V\_\{q-1\}^\wedge T\}\{|U\_q * S^\wedge\{1/2\}| * |V\_\{q-1\} * S^\wedge\{1/2\}|\},$$

where $Z(U,V)$ is the normalized factor.

It is possible to also apply word smoothing to the model-based K-Clustering as follows:

$$P(W\_q|d\_\{q\_1\}) = \backslash sum\_\{k=1\}^\wedge\{K\} P(W\_q|C\_k)P(C\_k|d\_\{q\_1\})$$

where $P(W\_q|C\_k)$, $P(C\_k|d\_\{q\_1\})$ can be computed using the distance measurement given above by a normalized factor.

In this way, the N-gram and LSA model are combined into a single language model and can be used for word comparison and text generation. The Python code for such LSA analysis was implemented in Thomo's [292] research.

In Table 94, we show statistical information on the semantically enhanced corpora produced previously and shown in Table 91.

Table 94. Specification of semantically generated corpora using LSA.

Data set	Number of Sentences		Number of unique Czech tokens	
	Monolingual	Generated	Monolingual	Generated
News 2007	100,766	98,726	200,830	72,975
News 2008	4,292,298	868,862	2,214,356	592,862
News 2009	4,432,383	895,127	2,172,580	729,972
News 2010	2,447,681	725,751	1,487,500	472,976
News 2011	8,746,448	1,197,762	2,871,190	829,927
News 2012	7,538,499	1,298,765	2,589,424	750,865
News 2013	8,886,151	1,314,276	2,768,010	694,290
News 2014	8,722,306	1,267,862	2,814,742	992,893
News 2015	8,234,140	1,471,287	2,624,473	892,291
TOTAL	53,366,020	9,138,418	19,743,105	6,029,051

5.4.3.4 *Experimental setup*

The machine translation experiments we conducted involved three WMT16 tasks: News translation, information technology (IT) document translation and biomedical text translation. Our experiments were conducted on the CS-EN pair in both directions. To obtain more accurate word alignment, we used the SyMGiza++ tool, which assisted in the formation of a similar word alignment model. This particular tool develops alignment models that obtain multiple many-to-one and one-to-many alignments in multiple directions between the given language pairs. SyMGiza++ is also used to create a pool of several processors, supported by the newest threading management, which makes it a very fast process. The alignment process used in our case utilizes four unique models during the training of the system to achieve refined and enhanced alignment outcomes. The results of these approaches have been shown to be fruitful in previous research [266]. OOV words are another challenge for an SMT system and to deal with such words, we used the Moses toolkit and the Unsupervised Transliteration Model (UTM). The UTM is a language-independent approach that has an unsubstantiated capability for learning OOV words. We also utilized the post-decoding transliteration method from this particular toolkit. UTM is known to make use of a transliteration phrase translation table to access probable solutions. UTM was used to score several possible transliterations and to find a translation table [267, 268, 79].

The KenLM tool was applied to language model training. This library helps to resolve typical problems of language models, reducing execution time and memory usage. To reorder the phrase probability, the lexical values of the sentences were used. We also used KenLM for lexical reordering. Three directional types are based on each target–swap (S), monotone (M), and discontinuous (D)–all three of which were used in a hierarchical model. The bidirectional restructuring model was used to examine the phrase arrangement probabilities [10, 267, 47].

The quality of domain adaptation largely depends on training data, which helps in incorporating the linguistic and translation models. The acquisition of domain-centric data helps greatly in this regard [113]. A parallel, generalized domain corpus and monolingual corpus were used in this process, as identified by Wang et al. [265]. First, sentence pairs of the parallel data were weighted based on their significance to the targeted domain. Second, reorganization was conducted to obtain the best sentence pairs. After obtaining the required sentence pairs, these models were trained for the target domain [265].

For similarity measurement, we used three approaches: Word overlap analysis, the cosine term frequency-inverse document frequency (tf-idf) criterion, and perplexity measurement. However, the third approach, which incorporates the best of the first two, is the strictest. Moreover, Wang et al. observed that a combination of these approaches provides the best possible solution for domain adaptation for Chinese-English corpora [265]. Thus, inspired by Wang et al.'s approach, we utilized a combination of these models. Similarly, the three measurements were combined for domain adaptation. Wang et al. found that the performance of this process yields approximately 20 percent of the domain analogous data.

5.4.3.5 Evaluation

Numerous human languages are used around the world and millions of translation systems have been introduced for the possible language pairs. However, these translation systems struggle with high quality performance, largely due to the limited availability of language resources, such as parallel data.

In this study, we have attempted to supplement these limited resources. Additional parallel corpora can be utilized to improve the quality and performance of linguistic resources, as well as individual NLP systems. In the MT application (Table 92), our data generation approach has increased translation performance. Although the results appear very promising, there remains a great deal of room for improvement. Performance improvements can be attained by applying more sophisticated algorithms in order to quantify the comparison among different MT engines. In Table 94, we present the baseline (BASE) outcomes for the MT systems we obtained for three diverse domains (news, IT, and biomedical—using official WMT16 test sets). Second, we generated a virtual corpus and adapted it to the domain (FINAL). The generated corpora demonstrate improvements in SMT quality and utility as NLP resources. From Table 91, it can be concluded that a generated virtual corpus is morphologically rich, which makes it acceptable as a linguistic resource. In addition, by retraining with a virtual corpus SMT system and repeating all the steps, it is possible to obtain more virtual data of higher quality. Statistically significant results in accordance with the Wilcoxon test are marked with * and those that are very significant with **.

Next, in Table 96, we replicate the same quality experiment but using generated data without the back-translation step. As shown in Table 92, more data can be obtained in such a manner. However, the SMT results are not as good as those obtained using back-translation. This means that the generated data must be noisy and most likely contain incomplete sentences that are removed after back-translation.

Next, in Table 97, we replicate the same quality experiment but using generated data from Table 93. As shown in Table 97, augmenting virtual

Table 95. Evaluation of generated corpora.

Domain	Direction	System	BLEU
News	CS→EN	BASE	15.26
	CS→EN	FINAL	18.11**
	EN→CS	BASE	11.64
	EN→CS	FINAL	13.43**
IT	CS→EN	BASE	12.86
	CS→EN	FINAL	14.12*
	EN→CS	BASE	10.19
	EN→CS	FINAL	11.87*
Bio-Medical	CS→EN	BASE	16.75
	CS→EN	FINAL	18.33**
	EN→CS	BASE	14.25
	EN→CS	FINAL	15.93*

Table 96. Evaluation of corpora generated without the back-translation step.

Domain	Direction	System	BLEU
News	CS→EN	BASE	15.26
	CS→EN	FINAL	17.32**
	EN→CS	BASE	11.64
	EN→CS	FINAL	12.73*
IT	CS→EN	BASE	12.86
	CS→EN	FINAL	13.52*
	EN→CS	BASE	10.19
	EN→CS	FINAL	10.74*
Bio-Medical	CS→EN	BASE	16.75
	CS→EN	FINAL	16.83*
	EN→CS	BASE	14.25
	EN→CS	FINAL	15.03**

corpora with semantic information makes a positive impact on not only the data volume but also data quality. Semantic relations improve the MT quality even more.

Finally, in Table 98, we replicate the same quality experiment but using generated data from Table 94 (LSA). As shown in Table 98, augmenting virtual corpora with semantic information by facilitating LSA makes an even more positive impact on data quality. LSA-based semantic relations improve the MT quality even more. It is worth mentioning that LSA provided us with

Table 97. Evaluation of semantically generated corpora without the back-translation step.

Domain	Direction	System	BLEU
News	CS→EN	BASE	15.26
	CS→EN	FINAL	19.31**
	EN→CS	BASE	11.64
	EN→CS	FINAL	14.87**
IT	CS→EN	BASE	12.86
	CS→EN	FINAL	15.42**
	EN→CS	BASE	10.19
	EN→CS	FINAL	12.17**
Bio-Medical	CS→EN	BASE	16.75
	CS→EN	FINAL	19.47**
	EN→CS	BASE	14.25
	EN→CS	FINAL	16.13**

Table 98. Evaluation of semantically generated corpora using LSA.

Domain	Direction	System	BLEU
News	CS→EN	BASE	15.26
	CS→EN	FINAL	19.87**
	EN→CS	BASE	11.64
	EN→CS	FINAL	15.61**
IT	CS→EN	BASE	12.86
	CS→EN	FINAL	16.18**
	EN→CS	BASE	10.19
	EN→CS	FINAL	13.04**
Bio-Medical	CS→EN	BASE	16.75
	CS→EN	FINAL	20.37**
	EN→CS	BASE	14.25
	EN→CS	FINAL	17.28**

less data but we believe that it was more accurate and more domain-specific than the data generated using Wordnet.

Summing up, in this study, we successfully built parallel corpora of satisfying quality from monolingual resources. This method is very time and cost effective and can be applied to any bilingual pair. In addition, it might prove very useful for rare and under-resourced languages. However, there is still room for improvement, for example, by using better alignment models, neural machine translation or adding more machine translation engines to our methodology. Moreover, using Framenet, which provides semantic roles for a word and shows restrictions in word usage, in that only several kinds of word can be followed by a certain word, might be of interest for future research [293].

5.4.4 *Statistical noisy parallel data filtration and adaptation*

Parallel sentences are a valuable source of information for machine translation systems and other cross-lingual information-dependent tasks. Unfortunately, such data is quite rare, especially for the Polish–English language pair. In general, the sentences of a parallel corpus must be aligned before using it for processing. Sentences in a raw corpus are sometimes misaligned, resulting in translated lines whose placement does not correspond to the text lines in the source language. Moreover, some sentences may have no corresponding translation in the corpus at all. The corpus might also contain poor or indirect translations, making alignment difficult and reducing the quality of data. Thus, corpora alignment is crucial for MT system accuracy [39]. This must also be computationally feasible in order to be of practical use in various applications [40].

Previous attempts to automatically compare sentences for parallel corpora were based on sentence lengths and vocabulary alignment [41]. Brown [43] proposed a method based on measuring the sentence length using the number of words. Gale and Church [42] measured the number of characters in sentences. Other studies explored methods that combine sentence length statistics with the alignment of vocabularies [23, 22]. A rule-based approach was proposed in [24]; however, its implementation and scalability for other languages is difficult. Nonetheless, such methods lead to the creation of noisy parallel corpora at best, or are not cost effective when applied to other languages.

On the other hand, the approach proposed in this study is language independent and cost- and time-effective. We used a statistical classifier to develop a filtering solution that can cope without. This method does not use any hard-coded rules, even for rare translation samples, and hence, can be easily adjusted for specific filtration needs or quality thresholds. Moreover, the proposed solution can be used as an in-domain adaptation technique, and further improved via bootstrapping. It is freely available on GitHub.

5.4.4.1 *Data acquisition*

In this research, the proposed method was applied to both noisy-parallel as well as comparable corpora. The former was obtained from the WIT3 [15] project (TED Talks), the latter was built from the Wikipedia articles.

Nonetheless, a high quality parallel corpus was required to build statistical classifiers. A similar corpus was also necessary to conduct in-domain adaptation experiments. We used the high quality human translated BTEC [17] corpus (tourist dialogs) for classifier training and very specific European Parliament Proceedings (EUP) [16] corpus for in-domain data adaptation. In addition, we also extended the BTEC corpus for some of the experiments to include additional data from movie subtitles. This was necessary in order to build a more extensive statistical classifier for quality comparison. This data was obtained from the OpenSubtitles (OPEN) corpora [294]. The corpora statistics are presented in the Table 99.

Table 99. Corpora statistics.

Corpus	Unique PL Tokens	Unique EN Tokens	Sentence Pairs
BTEC	50,782	24,662	220,730
TED	218,426	104,117	151,288
OPEN	1,236,088	749,300	33,570,553
EUP	311,654	136,597	632,565

The TED Talks were chosen as a representative sample of noisy parallel corpora. The Polish data in the TED talks (about 15 MB) included approximately 2 million words that were not tokenized. The transcripts themselves were provided as pure text encoded using the UTF-8 format [17].

Wikipedia data obtained in previous chapters was chosen as a comparable data source because of the large number of documents that it provides (1,047,423 articles in PL Wiki and 4,524,017 in EN, at the time of writing).

5.4.4.2 *Building bi-lingual phrase table*

The approach described in this research requires the construction of a bilingual translation table using the statistics gathered from the training corpus. Such a phrase table would allow us to evaluate whether a pair of sentences can be considered equivalent or not. Another advantage of this approach is that it is possible to generate an n-best list of most probable translations, prune the phrase table, and easily set likelihood thresholds [8].

The most common approach used to build the phrase table in an unsupervised manner is by analyzing large amounts of parallel texts developed by IBM and already discussed. In this study, we use an improved version of IBM Model 4. The tool SyMGiza++ [295] is used to create symmetric word alignment models. This tool generates alignment models that support many-to-one and one-to-many alignments in both directions between two language pairs. SyMGiza++ is also designed to leverage the power of multiple processors through advanced threading management, making it very fast. Its alignment process uses four different models during training to progressively refine alignment results. This approach has yielded impressive results, as discussed in [266].

5.4.4.3 Experiments and results

The solution presented in this study facilitates the construction of phrase tables [206] that allow us to determine the likelihood that a pair of sentences are translations of each other. As the approach is statistical in nature and the human language is full of inflections, synonym, hyponyms, etc., the tool not only checks the most likely translation, but also generates an n-best list of possible solutions. Both the probability threshold and the number of n-best results can be adjusted in order to restrict the results. Based on this information, a comparison is performed between the pair of sentences and the n-best list is generated. This approach allows the tool to evaluate hypothetical sentences [296].

The solution obtained using this procedure is highly dependent on the quality and quantity of the data used for phrase table training. Hence, we decided to prepare two tables, the first one from a smaller amount of data (BTEC corpus), and the second from a larger volume of data (BTEC corpus integrated with the OPEN corpus). Both these phrase tables were used in the experiments. Additionally 5,000 random sentence pairs were selected from the BTEC corpus for the purpose of evaluation. In order to cover the entire test file, we divided the BTEC corpus into 500 equal chunks and randomly selected 10 pairs from each chunk. The evaluation was conducted by measuring the influence of filtering on the quality of the machine translation systems. The Moses Statistical Machine Translation (SMT) toolkit was used for training the translation system [184].

Two systems were trained in each experiment. The first one used the TED corpus for training and the second one used the Wikipedia (WIKI) corpus for evaluation. The baseline system scores are indicated in the tables using the BASE suffix. For each experiment, the BLEU, NIST, TER and METOR metrics were used. The results after filtering using the smaller phrase table are denoted by TED-F and WIKI-F, while the results after filtering using the bigger phrase table are denoted by TED-FD and WIKI-FD. Table 100 presents the number of bi-sentences before and after filtration, while Table 101 shows the results of machine transition using those data sets. In addition, a statistical significance (Wilcoxon test) test was conducted in order to verify the changes in the metrics in comparison with the baseline systems. The results are presented in Table 101. Significant changes are indicated using a * and very significant changes are marked as **.

Table 100. Data amount after filtration.

Corpus	Number of Bi-sentences
TED BASE	151,288
TED-F	76,262
TED-FD	109,296
WIKI BASE	475,470
WIKI-F	153,526
WIKI-FD	221,732

Table 101. SMT results after data filtration.

System	BLEU	NIST	TER	METEOR
TED BASE	23.54	6.64	53.84	52.42
TED-F*	22.17	6.45	54.22	52.13
TED-FD**	25.74	6.97	51.54	54.36
WIKI BASE	16.97	5.24	64.23	49.23
WIKI-F**	18.43	5.67	59.52	50.97
WIKI-FD**	19.18	6.01	55.26	51.35

Phrase tables often grow enormously large, because they contain a lot of noisy data and store many unlikely hypotheses. This poses difficulties when implementing a real-time translation system that has to be loaded into memory. It might also result in the loss of quality. Quirk and Menezes [123] argue that extracting only minimal phrases (the smallest phrase alignments, each of which map to an entire sentence pair) will negatively affect translation quality. This idea is also the basis of the n-gram translation model [124, 125]. On the other hand, the authors of [126] postulate that discarding unlikely phrase pairs based on significance tests drastically reduces the phrase table size and might even result in performance improvement. Wu and Wang [127] propose a method for filtering the noise in the phrase translation table based on a logarithmic likelihood ratio. Kutsumi et al. [128] use a support vector machine for cleaning phrase tables. Eck et al. [129] suggest pruning of the translation table based on how often a phrase pair occurred during the decoding step and how often it was used in the best translation. In this study, we implemented the Moses-based method introduced by Johnson et al. in [126].

In our experiments, absolute and relative filtration was performed based on in-domain dictionary filtration rules. Different rules were used for the sake of comparison. For example, one rule (absolute) was to retain a sentence if at least a minimum number of words from the dictionary appeared in it. A second rule (relative) was to retain a sentence if at least a minimum percentage of words from the dictionary appeared in it. A third rule was to retain only a set number of the most probable translations. Finally, the experiment was concluded using a combination of pruning and absolute pre-filtering.

In Table 102, results are provided for the experiment that implemented the Moses-based pruning method, along with data pre-processing performed by filtering irrelevant sentences. The absolute and relative filtration results are indicated using the following terminology.

Absolute N—Retain a sentence if at least N words from the dictionary appear in it.

Relative N—Retain a sentence if at least N% of the sentence is built from the dictionary.

Pruning N—Remove only the N least probable translations.

The Table 102 also contains results for the significance tests for the pruning experiments.

Table 102. Pruning results.

Pruning Method	Phrase Table	BLEU	NIST	TER	METEOR
Baseline	TED	25.74	6.97	51.54	54.36
Absolute 1*	TED	26.11	7.01	51.23	54.74
Absolute 2*	TED	24.53	6.41	52.64	53.17
Relative 1*	TED	25.14	6.91	51.74	53.78
Relative 2**	TED	23.34	5.85	54.13	51.83
Pruning 20**	TED	27.51	7.28	46.32	58.92
Pruning 40	TED	25.22	6.39	53.65	52.51
Baseline	WIKI	19.18	6.01	55.26	51.35
Absolute 1	WIKI	19.02	5.88	55.71	50.83
Absolute 2*	WIKI	20.74	6.03	54.37	52.21
Relative 1**	WIKI	18.55	5.89	55.73	50.93
Relative 2**	WIKI	17.42	4.73	61.64	47.82
Pruning 20**	WIKI	21.73	6.58	50.62	54.89
Pruning 40*	WIKI	21.26	6.27	51.44	53.79

The results were positive in terms of the machine translation quality. We decided to improve it further using the bootstrapping method. As the BTEC-based phrase table was smaller, we used it in our baseline filtration experiment. The filtering was conducted on the WIKI corpus because it contained more data. First, the WIKI corpus was filtered using the baseline phrase table. Then, the newly obtained data was added to the BTEC corpus and the phrase table was retrained. Finally, the new phrase table was used to filter the WIKI corpus again. This process was repeated iteratively 10 times. After each iteration, the SMT system was trained on the resulting WIKI-F corpus. As seen in Table 103, each iteration slightly improved the results by reducing the data loss. The first few iterations showed more rapid improvement. The TED corpus was used as a baseline system.

Table 103. Pruning impact on SMT quality.

Iteration	Obtained Bi-sentences	BLEU	NIST	TER	METEOR
Baseline	151,288	23.54	6.64	53.84	52.42
1**	76,262	22.17	6.45	54.22	52.13
2*	89,631	22.94	6.51	54.14	52.31
3*	94,827	24.11	6.79	53.12	54.01
4**	98,131	25.21	6.88	51.32	55.46
5**	101,003	25.89	6.92	50.49	56.22
6**	102,239	26.07	7.02	49.56	56.73
7*	102,581	26.12	7.04	49.51	56.74
8	102,721	26.11	7.04	49.52	56.69
9	102,784	25.53	6.97	51.28	55.93
10	102,853	25.18	6.82	51.38	55,79

Finally, the method proposed in this research was evaluated as a data domain adaptation technique. It was compared with other commonly used methods by measuring the influence on the machine translation quality. The methods used in the comparison were: The in-built Moore-Levis filtering (MML) method, Levenshtein Distance (LEV), Cosine TF-IDF (COS), and Perplexity (per). The proceedings of the European Parliament (EUP) corpus were taken as the baseline for training the SMT system. Its domain is limited, so the phrase table for filtering was also trained using this data set (EUP-F). Using the abovementioned methods, the WIKI corpus was adapted by in-domain data selection. Moreover, the SMT system was retrained using EUP data and the procedure was carried out. The evaluation results, along with the significance tests, are showed in Table 104.

Table 104. Data adaptation evaluation.

Adaptation Method	BLEU	NIST	TER	METEOR
None	57.36	9.43	32.76	73.58
MML**	59.25	9.89	27.46	75.43
LEV*	58.45	9.32	29.62	73.81
COS**	58.78	9.47	28.31	75.12
PER**	58.11	8.43	28.44	75.32
EUP-F**	58.74	8.44	27.85	75.39

In general, it is difficult to create high quality parallel text corpora. Sentences in a raw corpus might suffer from issues such as misalignment, lack of corresponding translation, and poor or indirect translations, resulting in a poor quality of data. In this study, we proposed a technique to eliminate incorrect translations and noisy data from parallel and comparable corpora, with the ultimate goal of improving SMT performance. The experimental evaluation of the proposed method showed that it is effective in filtering noisy data and improving the SMT quality by a significant factor. Further improvement using bootstrapping or phrase table pruning was also demonstrated. The proposed technique can also be used for in-domain data adaptation. The evaluation did not provide the best results but we plan on improving its performance in the future. The proposed approach is also language independent and can be adapted for use in processing positional and lexically similar languages by simply adjusting the initial parameters and training data. We demonstrated this by using it for English and Polish processing. From a practical point of view, the method requires neither expensive training nor language-specific grammatical resources, while producing satisfactory results. In our future work, we plan to improve this method by incorporating more advanced phrase table training models like IBM Model 5 or Model 6. They should be computationally feasible if implemented using GPU. We also plan to explore neural networks for this task [297].

5.4.5 Mixing textual data selection methods for improved in-domain adaptation

The performance of statistical machine translation (SMT) [182] is heavily dependent on the quantity of training data and the domain specificity of the test data as it relates to the training data. An obstacle to optimal performance is the data-driven system not guaranteeing optimum results if either the training data or testing data are not uniformly distributed. Domain adaptation is a promising approach to increasing the quality of domain-specific translation systems with a mixture of out-of-domain and in-domain data.

The prevalent adaptation method is to choose the data to target the field or domain from a general cluster of documents within the domain. However, the method is applied when the quantity of data is adequately wide to cover some sentences that will exist in the targeted field. Moreover, a domain-adapted machine translation system can be attained through training that uses a chosen subset of the data. Axelrod et al. [113] explained this point as a quasi-in-domain sub-part of the corpus instead of using the complete corpus data.

The present research focuses entirely on these auxiliary data selection methodologies, which have advanced the development of narrow-domain SMT systems. Translation models that are trained in this way will have the benefit of enhanced word arrangements. Furthermore, the system can be modified to avoid redundant pairs of phrases; moreover, proper estimation can support reorganization of the elements of target sentences.

The similarity measurement has an immense influence on the translation quality. In this study, data selection methods that can boost the quality of domain-specific translation were explored. To this end, data selection criteria were carefully analyzed. Two models are, thus, considered in this monograph. One is based on cosine term frequency–inverse document frequency (tf-idf); the other is a perplexity-based approach. These two techniques have roots in information retrieval (IR) and language modeling for SMT. A third approach uses the Levenshtein distance, which is then analyzed.

An evaluation revealed that each of these methods has advantages and disadvantages. First, the cosine tf-idf technique employs the text as a set of words and recovers sentences that are similar. Although this approach helps reduce the number of out-of-vocabulary words, it does not filter bad data. On the other hand, perplexity-oriented measurement tools leverage an n-gram language model (LM), which considers both a grammar's word order and term distribution. It filters irrelevant phrases and out-of-vocabulary words. However, the quality of the filtering depends largely on the in-domain LM and quasi-in-domain sub-parts.

The methodology based on the Levenshtein distance is more stringent in its approach than the other two. It is intended to explore the similarity index; however, in terms of performance, it does not surpass the others on account of its reliance on data generalization. As the number of factors increases, the complexity of the similarity judgment also increases. This scenario can be depicted by a pyramid to show the relevant intensities of multiple approaches. The Levenshtein distance approach is at the top of the pyramid, perplexity is in the middle, and the cosine tf-idf approach is at the bottom. The positive and negative aspects of each method can be addressed by considering all these criteria. If we consider additional factors, then the criterion at each highest point will become stricter.

In this study, the above measurement approaches were combined and compared to each separate method and to the modified Moore–Lewis filtering implementation in the Moses SMT system. A comparative experiment was conducted using a generalized Polish–English language movie-subtitle corpus and in-domain TED lecture corpus. The SMT systems were adapted and trained accordingly. Utilizing the bilingual evaluation understudy (BLEU) metric, the testing results revealed that the designed approach produced a promising performance.

Existing literature discusses data adaptation for SMT from multiple perspectives, such as finding unknown words from comparable corpora [298], corpora weighting [82], mixing multiple models [107, 106] and weighted phrase extraction. The predominant criterion for data selection is cosine tf-idf, which originated in the area of IR. Hildebrand et al. [102] utilized this IR technique to choose the most similar sentence—albeit with a lower quantity—for translation model (TM) and LM adaptation. The results strengthen the significance of the methodology for enhancing translation quality, particularly for LM adaptation.

In a study much closer to the present research, Lü et al. [105] suggested reorganizing the methodology for offline, as well as online, TM optimization. The results are much closer to those of a realistic SMT system. Moreover, their conclusions revealed that repetitive sentences in the data can affect the translation quality. By utilizing approximately 60% of the complete data, they increased the BLEU score by almost one point.

The second technique in the literature is a perplexity approach, which is common in language modeling. This approach was used by Lin et al. [299] and Gao et al. [300]. In that research, perplexity was utilized as a standard in testing parts of the text in accordance with an in-domain LM approach. Other researchers, such as Moore and Lewis [269], derived the unique approach of a cross-entropy difference metric from a simpler version of the Bayes rule. This methodology was further examined by Axelrod et al. [113], particularly for SMT adaptation, and they additionally introduced an exclusive unique bilingual methodology and compared its results with contemporary approaches. Results of their experiments revealed that, if the system was kept simple yet sufficiently fast, it discarded as much as 99% of the general corpus, which resulted in an improvement of almost 1.8 BLEU points.

Early works discuss separately applying the methodology to either a TM [113] or an LM [105]; however, in [105], Lü suggests that a combination of LM and TM adaptation will actually enhance the overall performance. Therefore, in the present study, TM and LM optimization was investigated through a combined data selection method.

5.4.5.1 *Combined corpora adaptation method*

Four selection criteria are discussed to describe the examined models: cosine tf-idf, perplexity, Levenshtein distance, and the proposed combination approach.

5.4.5.1.1 Cosine tf-idf

In the approach based on cosine tf-idf, each document D_i is represented as a vector $(w_{i1}, w_{i2}, ..., w_{in})$, where n is the vocabulary size. Thus, w_{ij} is calculated as:

$$w_{ij} = tf_{ij} \times \log(idf_j)$$

where $tf_{ij}i$ is the term frequency (TF) of the j-th word in the vocabulary in document D_i, and idf_j is the inverse document frequency (IDF) of the j-th word. The similarity between the two texts is the cosine of the angle between the two vectors. This formula is applied in accordance with Lü et al. [105] and Hildebrand et al. [102]. The approach supposes that M is the size of the query set and N is the number of similar sentences from the general corpus for each query. Thus, the size of the cosine tf-idf-based quasi-in-domain sub-corpus is defined as:

$$Size_{Cos-IR} = M \times N$$

5.4.5.1.2 Perplexity

Perplexity focuses on cross-entropy [301], which is the average of the negative logarithm of word probabilities. Consider:

$$H(p,q) = -\sum_{i=1}^{n} p(w_i) \log q(w_i) = -\frac{1}{N}\sum_{i=1}^{n} \log q(w_i)$$

where p is the empirical distribution of the test sample. If w_i appears n times in the test sample of size N, then $q(w_i)$ is the probability of the w_i event approximated from the training set.

Perplexity (pp) can be simply calculated at the base point presented in the system. It is often applied as a symbolic alternative to perplexity for the data selection as:

$$pp = b^{H(p, q)}$$

where b is the basis of measured cross-entropy, and $H(p, q)$ is the cross-entropy as given in [301] (which is often used as a substitute for perplexity in data selection [113, 269]).

Let $H_I(p, q)$ and $H_O(p, q)$ be the cross-entropy of the w_i string in accordance with the language model, which is subsequently trained by a general-domain dataset and an in-domain dataset. While examining the target (tgt) and source (src) dimensions of the training data, three perplexity-based variants exist. The first one, known as basic cross-entropy, is defined as:

$$H_{I-src}(p, q)$$

The second is the Moore–Lewis cross-entropy difference (Moore and Lewis 2010),

$$H_{I-src}(p, q) - H_{G-src}(p, q),$$

which attempts to choose the sentences that are most identical to the ones in I but unlike others in G. Both the standards mentioned above consider only sentences in the source language. Moreover, Axelrod et al. [113] proposed a metric that adds cross-entropy differences to both sides:

$$[H_{I-src}(p, q) - H_{G-src}(p, q)] + [H_{I-tgt}(p, q) - H_{G-tgt}(p, q)]$$

For instance, candidates with lower scores [298, 27] have a higher relevance to the specific target domain. The sizes of the perplexity-based quasi-in-domain subsets must be equal. In practice, we work with the SRI Language Modeling (SRILM) toolkit to train 5-gram LMs with interpolated modified Kneser–Ney discounting [9, 32].

5.4.5.1.3 Levenshtein distance

In information theory and computer science, the Levenshtein distance is regarded as a string metric for the measurement of dissimilarity between two sequences. The Levenshtein distance between points or words is the minimum possible number of unique edits to the data (e.g., insertions or deletions) that are required to replace one word with another.

The Levenshtein distance can additionally be applied to a wider range of subjects as a distance metric. Moreover, it has a close association with pairwise string arrangement.

Mathematically, the Levenshtein distance between two strings a, b (of length $|a|$ and $|b|$, respectively) is given by $lev_{a, b}(|a|, |b|)$, where

$$lev_{a,b}(i,j) = \begin{cases} \max(i,j) & if \min(i,j) = 0 \\ \min \begin{cases} lev_{a,b}(i-1,j) + 1 \\ lev_{a,b}(i,j-1) + 1 \\ lev_{a,b}(i-1,j-1) + 1_{(a_i \neq b_j)} \end{cases} & otherwise. \end{cases}$$

Here, $1_{(a_i \neq b_j)}$ is the indicator function, which is equal to 0 when $a_i = b_j$; otherwise, it is equal to 1. Furthermore, $lev_{a,b}(i, j)$ is the distance between the first i characters of a and the first j characters of b. The first component has the least correspondence to the deletion (from a to b), the second-closest correspondence to the insertion, and the most correspondence to a match or mismatch.

5.4.5.1.4 Combined methods[49]

As was first explained by Wang et al. [265], there are three basic processing stages in data selection for domain adaptation. First, we extract sentence pairs from a parallel corpus. A generalized domain corpus is obtained based on

---

[49] https://github.com/krzwolk/Text-Corpora-Adaptation-Tool

significance and corresponding relevance to the targeted domain. Second, the samples are reorganized in order to maintain the quasi-in-domain sub-corpus. These first two steps are applicable to a general domain monolingual corpus and they are significant for selecting sentences for a language model. Once a large number of sentence pairs are collected, these models are scheduled for data training and will eventually represent the target domain.

In a similar fashion, the similarity index measurement is required in order to choose the sentences for a quasi-in-domain sub-corpus. For the similarity measurement, three approaches are regarded as the most suitable. First, the cosine tf-idf criterion identifies the similarity by considering the word overlap. This technique is particularly helpful in reducing out-of-vocabulary words. Nevertheless, it is sensitive to irrelevant data in the system. The perplexity-based criterion, on the other hand, is more focused on the n-gram word order. Meanwhile, the Levenshtein distance considers the word order, position of the words and word overlap. Of the three approaches, it is the most stringent.

In this study, a combination of the corpora and language models is used. These three methods are first individually used to identify the quasi-in-domain sub-corpora. They are later combined during the reorganization phase to collectively leverage the benefits of all three metrics. Similarly, these three metrics are joined for domain adaptation during the translation process. Experimental evidence demonstrated the success of this process. In addition, our adaptation tool is freely available for use.[50]

5.4.5.2 *Results and conclusions*

TED data comprise a unique lecture domain; however, this domain is not as wide as that of the movie subtitles corpus OpenSubtitles (OPEN). An SMT system most effectively operates in a uniquely defined domain, which presents another challenge for the system. If the challenge is not adequately addressed, it can decrease the translation accuracy. The domain adaptation quality largely depends on the training data used to optimize the language and translation models in the SMT system. This challenge can be addressed by selecting and extracting domain-centric training data from a general corpus and a generalized domain monolingual corpus. The quasi-in-domain sub-corpus is produced through this process.

In this study, experiments were conducted on the Polish–English language pair. The corpora statistics are shown in Table 104. In the Polish (PL) and English (EN) columns, the number of unique words is given for each language; the number of bilingual sentences is given in the "PAIRS" column.

---

[50] https://github.com/krzwolk/Text-Corpora-Adaptation-Tool

Table 105. Corpora statistics.

CORPORA	PL	EN	PAIRS
TED	218,426	104,117	151,228
OPEN	1,236,088	749,300	33,570,553

In Table 105, the corpora statistics are presented for the average sentence lengths for each language and corpus. Both tables expose large disparities between the text domains.

Table 106. Average sentence lengths.

CORPORA	PL	EN
TED	13	17
OPEN	6	7

Multiple versions of the SMT system were evaluated through the experiments. Using the Moses SMT system, we trained a baseline system with no additional data (BASE), a system that employs additional subtitle corpora with no adaptation (NONE), a system adapted using Moore–Lewis filtering (MML) [113] built into Moses, a system using tf-idf adaptation (TF-IDF), a system using perplexity-based adaptation (PP), a system using data selected by the Levenshtein distance (LEV), and, lastly, a system combining the three methods as described in Section 3.4 (COMB). In Table 106, we present the amount of data from the OPEN corpus that remained after each filtration method.

Table 107. Number of remaining bi-sentences after filtration.

Filtration method	Number of bi-sentences
NONE	33,570,553
MML	1,320,385
TF-IDF	1,718,231
PP	2,473,735
LEV	1,612,946
COMB	983,271

Additional data were used for training both the bilingual translation phrase tables and language models. The Moses SMT system was used for tokenization, cleaning, factorization, conversion to lower case, splitting, and final cleaning of corpora after splitting. Training of a 6-gram language model was accomplished using the KenLM Modeling Toolkit [9]. Word and phrase alignment was performed using the SyMGIZA++ tool [266].

Out-of-Vocabulary (OOV) words were addressed using an unsupervised transliteration model [79]. For evaluation purposes, we used an automatically calculated BLEU metric [18] and official International Workshop on Spoken Language Translation (IWSLT) 2012 test datasets.[51] The results are shown in Table 107. Statistically significant results in accordance with the Wilcoxon test are marked with an asterisk '*'; those that are very significant are denoted with '**.'

Table 108. Corpora adaptation results.

SYSTEM	BLEU	
	PL->EN	EN->PL
BASE	17.43	10.70
NONE	17.89*	10.63*
MML	18.21**	11.13*
TF-IDF	17.92*	10.71
PP	18.13**	10.88*
LEV	17.66*	10.63*
COMB	18.97**	11.84**

As shown by Table 107, ignoring the adaptation step only slightly improves PL->EN translation and degrades EN<-PL translation. As anticipated, other adaptation methods have a rather positive impact on translation quality; however, in some cases, the enhancement is only minor.

The most significant improvement in translation quality was obtained using the proposed method combining all three metrics. It should be noted, however, that the proposed method was not computationally feasible in some cases, even though it produced satisfactory results. In the best-case scenario, fast comparison metrics, such as perplexity, will filter most irrelevant data; however, in the worst-case scenario, most data would be processed by slow metrics.

Summing up, we successfully introduced a new combined approach for the in-domain data adaptation task. In the general case, it provides better adaptation results than those of state of the art methods separately in a reasonable amount of time.

5.4.6 *Improving ASR translation quality by de-normalization*

In natural spoken language there are many meaningless modal particles and dittographes. Furthermore, ASR often produces some recognition errors, and

---

[51] iwslt.org

the ASR results have no punctuation or proper casing. Therefore, the translation would be rather poor if the ASR results are directly translated by MT systems. Automatic sentence segmentation of speech is important in making speech recognition (ASR) output more readable and easier for downstream language processing modules. Various techniques have been studied for automatic sentence boundary detection in speech, including hidden Markov models (HMMs), maximum entropy, neural networks and Gaussian mixture models, utilizing both textual and prosodic information.

This part of research addresses the problem of identifying sentence boundaries in the transcriptions produced by automatic speech recognition systems. These differ from the sorts of texts normally used in NLP in a number of ways: The text is generally in single case, unpunctuated, and may contain transcription errors. In addition, such text is usually normalized, what is unnatural to read. A lot of important linguistic information is lost as well (e.g., for machine translation or text-to-speech). Table 108 compares a short text in the format that would be produced by an ASR system with a fully punctuated version that includes casing information and de-normalization.

There are many possible situations in which an NLP system may be required to process ASR text. The most obvious examples are text to speech systems, machine translation or real-time speech-to-speech translation systems. Dictation software programs do not punctuate or capitalize their output. However, if this information could be added to ASR text, the results would be far more useful.

One of the most important pieces of information that is not available in ASR output is sentence boundary information. The knowledge of sentence boundaries is required by many NLP technologies. Parts of speech taggers typically require input in the format of a single sentence per line, and parsers generally aim to produce a tree spanning each sentence. Sentence boundary is very important in order not to lose the context or even the meaning of a sentence. Only the most trivial linguistic analysis can be carried out on text that is not split into sentences.

It is worth mentioning that not all transcribed speech can be sensibly divided into sentences. It has been argued by Gotoh and Renals [302], that the main unit in spoken language is the phrase, rather than the sentence. However, there are situations in which it is appropriate to consider spoken language to be composed of sentences. The need for such tools is also well-known in other NLP branches. For example, the spoken portion of the British National Corpus [303] contains 10 million words and was manually marked with sentence boundaries. A technology that identifies sentence boundaries could be used to speed up such processes.

Table 109. Sample input and output.

Before ile kobiet mających czterdzieści cztery lata wygląda tak jak amerykańska gwiazda na okładce swej nowej płyty jennifer lopez nie bierze jednak udziału w konkursie piękności czy więc również mocno jak do rzeźbienia swego ciała przyłożyła się do stworzenia dobrych piosenek ostatnie lata nie były dobre dla latynoskiej piosenkarki między innymi rozwody i rozstania zdewastowały jej poczucie kobiecej wartości a klapy ostatnich albumow o mały włos nie zrujnowały piosenkarskiej kariery
After Ile kobiet mających 44 lata wygląda tak, jak amerykańska gwiazda na okładce swej nowej płyty Jennifer Lopez? Nie bierze jednak udziału w konkursie piękności. Czy więc również mocno, jak do rzeźbienia swego ciała, przyłożyła się do stworzenia dobrych piosenek?. Ostatnie lata nie były dobre dla latynoskiej piosenkarki, m.in. rozwody i rozstania zdewastowały jej poczucie kobiecej wartości, a klapy ostatnich albumow o mały włos nie zrujnowały piosenkarskiej kariery.

It is important to distinguish the aforementioned problem from another problem, sometimes called "sentence splitting". This problem aims to identify sentence boundaries in standard text. However, since standard text includes punctuation, the problem is effectively reduced to deciding which of the symbols that potentially denotes sentence boundaries (., !, ?) actually do. This problem is not trivial since these punctuation symbols do not always occur at the end of sentences. For example, in the sentence "Mr. Doe teacher at M.I.T." only the final full stop denotes the end of a sentence. For the sake of clarity, we shall refer to the process of discovering sentence boundaries in standard punctuated text as "punctuation disambiguation" and that of finding them in unpunctuated ASR text as "sentence boundary detection".

Sentence boundary detection is not all that is needed. We cannot forget about restoring proper casing, distinguishing between normal sentences and questions, or exclamatory sentences. Restoring numbers, dates, separators, etc. is not a trivial task, so it is still important to also obtain linguistic information from other tools.

5.4.6.1 De-Normalization method

In order to achieve this task, we implement a tool that works in a specialized pipeline of minor tools. We wanted it to be as easy for non-IT specialists as possible. In the first step, we used a conditional random fields framework for building probabilistic models in order to segment and label sequential data. Conditional random fields in this scenario offered several advantages over other methods such as hidden Markov models and stochastic grammars. CRFs include the ability to relax strong independence assumptions made in the other models. They also avoid a fundamental limitation of maximum entropy Markov models and other discriminative Markov models, which can be biased towards states with few successor states. A CRF differs from the HMM with respect to its training objective function (joint versus conditional likelihood) and its handling of dependent word features. Traditional HMM training does not maximize the posterior probabilities of the correct labels; whereas, the CRF directly estimates posterior boundary label probabilities $P(E|O)$. The underlying n-gram sequence model of an HMM does not cope well with multiple representations of a word sequence, especially when the training set is small. On the other hand, the CRF model supports simultaneously correlated features, providing more flexibility to incorporate a variety of knowledge sources. The CRFSuite is used for sentence boundary detection, as well as for comma separation detection.

For tasks such as conversion of numbers or capitalization, we created specialized rules. A specific python tool was implemented for this job. It required us to prepare large databases that were obtained using web crawlers, manually cleaned and annotated. It was necessary to annotate, for example, surnames that could be either casual words or names. The same work was required for enumeration, abbreviations, numerals, currency, etc. The surnames database consisted of 399,551 records, 2769 names, and 171 abbreviations. We also created a database of special cases for the Polish language. In the program, we assured the ease of extending rules and the databases to support future research.

5.4.6.2 *De-Normalization experiments*

In our experiments, we focused as much as possible on natural human speech in real-life scenarios. In order to do that, we chose three different types of data to process. One of the first corpora, similar to human dialogues, that we used was the Open Subtitles corpora obtained from the OPUS project9. In addition, we obtained multi-subject TED Lectures from the IWSLT'13 campaign proceedings and from the European Parliament Proceedings, whose subject domain vocabulary is very limited.

The corpora were randomly divided into training, tuning, and testing data. For training and tuning, we selected 1000 sentences for each task. In order to maximize the relevance of evaluation and coverage of the entire corpora, our randomization scripts first divided the corpora into 250 equal segments and then randomly chose four sentences from each segment. The selected sentences were removed from the training data.

Table 110 represents both the data specification and results of our initial sentence boundary detection script. The table rows represent: The number of sentences in each corpora, the number of words, the number of unique words and their forms, the average sentence length (words), and the measured accuracy of the trained system.

Table 110. Results of the tool.

	TED	EuroParl	OpenSubtitles
Sentences	185,637	629,558	9,745,212
Words	2,5 M	28 M	113 M
Unique Words	92,135	311,186	1,519,771
Avg. Sen. Len.	5	21	13
Accuracy	43%	67%	71%

From each corpus we randomly selected texts for development and testing, 500 lines for each purpose. These lines were deleted from the corpora for more reliable evaluation. With our translation system we processed plain ASR output from randomly selected lines (PLAIN), ASR output processed with our de-normalization tool (TOOL) and ASR output processed by a human translator (HUMAN). The results are showed in Table 111.

Table 111. Results of MT.

	Experiment	BLEU	NIST	MET	TER
TED	PLAIN	9.31	4.38	43.37	77.15
	TOOL	11.04	4.57	45.12	69.15
	HUMAN	16.05	5.34	49.24	66.42
EUROPARL	PLAIN	31.13	8.14	69.69	50.69
	TOOL	41.24	8.98	73.16	46.19
	HUMAN	63.18	10.78	82.05	33.14
OPENSUB	PLAIN	22.78	4.78	48.12	69.01
	TOOL	32.88	6.69	52.18	54.17
	HUMAN	53.21	7.57	66.40	46.01

This research has introduced sentence boundary detection on the text produced by the ASR systems as an area for application of the NLP technology. It is not possible to state whether the boundary detection works best for any specific data set, because of the variety and amount of data. It is clear that significant accuracy was obtained on a very narrow domain data, in which many phrases were repeated, even though this data was built from rather long sentences. In our opinion, the most valuable results are those of the subtitles corpora, because their sentence length is average and the data is diverse. It must be noted that we consider accuracy results over 60% as satisfactory, because we noticed that even humans, given ASR results, are not able to reproduce original text with 100% accuracy. In the future, combining training data and using additional Polish corpora in the training phase will most likely improve the accuracy of our de-normalization system. On the basis of manual analysis of the test samples, we concluded that additional changes to and extension of the implemented rules are also required.

The preliminary results of the machine translations are promising. Our experiments showed that overcoming outstanding issues, such as the lack of sentence boundaries, punctuation, and capitalization in the ASR output (generally a stream of words that lack sentence boundaries, which breaks the SMT process), improves the quality of MT systems. The automatic solution is still not as good as humans, but opportunities for improvement remain.

6
Final Conclusions

The main topic of this monograph stated that it is possible to improve spoken language translation quality by enriching textual training data with comparable corpora. Building and exploring comparable corpora in search for parallel data fulfilled this monograph. Moreover, it was proven that training parameters must be adapted for each language and text domain independently. It was also shown that improvements to corpora quality are essential to spoken language statistical machine transition. This confirms Monograph 2 of this monograph. To be more precise, SMT systems were trained, and their quality was improved in comparison to baseline systems. The best settings for PL-EN translation were determined for TED lectures data and evaluated during the IWSLT evaluation campaign, in which the best results for the PL-EN language pair were obtained twice in a row (2013 and 2014 evaluations). More importantly, the progress made for the PL-EN was one of the most significant, compared to the baseline.

Two methods for comparable corpora exploration were developed and tested. The Yalign tool was adapted and greatly improved, and almost 500,000 bi-sentences were successfully obtained from Wikipedia. The obtained data proved to be of acceptable quality and also improved the quality of the SMT systems. A second methodology for mining the bi-sentences was investigated, and its results look promising and worth additional research and development. In addition to those main tasks, a tool that can filter and align textual data was developed and proved to provide reasonable quality within this research effort. A successful trial was made to create an evaluation metric more suited for morphologically-rich languages. Many pre-processing tools were implemented and used as well (e.g., text cleaner applied to TED corpus, crawlers, text unifiers, etc.). All the solutions developed in the scope of this research are language independent and can easily be adapted for other languages, as well as for other text domains.

There is still room for improvement and research in statistical machine translation, and neural machine translation was recently proposed as an approach to MT. Unlike traditional statistical machine translation, neural machine translation aims at building a single neural network that can be jointly tuned to maximize translation performance. The models recently proposed for neural machine translation belong to a family of encoder-decoders. They consist of an encoder that encodes a source sentence into a fixed-length vector, from which a decoder generates a translation. Testing and improving such a methodology on the PL-EN language pair in accordance with [78] seems like a promising avenue for future research. Additionally, supplying SMT systems with neural-based language models has already proven to be a quality-improving approach [78, 131, 132]. What is more, plain neural-based translation systems recently began developing rapidly [78].

References

1. Mohammadi, Mehdi and Ghasemaghaee, Nasser. 2010. Building bilingual parallel corpora based on wikipedia. *In*: Computer Engineering and Applications (ICCEA), 2010 Second International Conference on. IEEE, pp. 264–268.
2. Adafre, Sisay Fissaha and De Rijke, Maarten. 2006. Finding similar sentences across multiple languages in wikipedia. *In*: Proceedings of the 11th Conference of the European Chapter of the Association for Computational Linguistics, pp. 62–69.
3. Yasuda, Keiji and Sumita, Eiichiro. 2008. Method for building sentence-aligned corpus from wikipedia. *In*: 2008 AAAI Workshop on Wikipedia and Artificial Intelligence (WikiAI08).
4. Tyers, Francis M. and Pienaar, Jacques A. 2008. Extracting bilingual word pairs from Wikipedia. Collaboration: Interoperability between people in the creation of language resources for less-resourced languages, p. 19.
5. Pal, Santanu, Pakray, Partha and Naskar, Sudip Kumar. 2014. Automatic building and using parallel resources for SMT from comparable corpora. *In*: Proceedings of the 3rd Workshop on Hybrid Approaches to Translation (HyTra)@ EACL, pp. 48–57.
6. Plamada, Magdalena and Volk, Martin. 2013. Mining for domain-specific parallel texts from the Wikipedia. Proceedings of the Sixth Workshop on Building and Using Comparable Corpora, pp. 112–120.
7. Smith, Jason R., Quirk, Chris and Toutanova, Kristina. 2010. Extracting parallel sentences from comparable corpora using document level alignment. *In*: Human Language Technologies: The 2010 Annual Conference of the North American Chapter of the Association for Computational Linguistics. Association for Computational Linguistics, pp. 403–411.
8. Koehn, Philipp et al. 2007. Moses: Open source toolkit for statistical machine translation. *In*: Proceedings of the 45th annual meeting of the ACL on interactive poster and demonstration sessions. Association for Computational Linguistics, pp. 177–180.
9. Stolcke, Andreas et al. 2002. SRILM-an extensible language modeling toolkit. *In*: Interspeech.
10. Heafield, Kenneth. 2011. KenLM: Faster and smaller language model queries. *In*: Proceedings of the Sixth Workshop on Statistical Machine Translation. Association for Computational Linguistics, pp. 187–197.
11. Gao, Qin and Vogel, Stephan. 2008. Parallel implementations of word alignment tool. *In*: Software Engineering, Testing, and Quality Assurance for Natural Language Processing. Association for Computational Linguistics, pp. 49–57.
12. Huang, Chu-Ren et al. 2007. Rethinking Chinese word segmentation: Tokenization, character classification, or word break identification. *In*: Proceedings of the 45th Annual Meeting of the ACL on Interactive Poster and Demonstration Sessions. Association for Computational Linguistics, pp. 69–72.

13. Wu, Dekai and Fung, Pascale. 2005. Inversion transduction grammar constraints for mining parallel sentences from quasi-comparable corpora. *In*: Natural Language Processing–IJCNLP 2005. Springer Berlin Heidelberg, pp. 257–268.

14. Tiedemann, Jörg. 2012. Parallel data, tools and interfaces in OPUS. *In*: LREC, pp. 2214–2218.

15. Cettolo, Mauro, Girardi, Christian and Federico, Marcello. 2012. Wit3: Web inventory of transcribed and translated talks. *In*: Proceedings of the 16th Conference of the European Association for Machine Translation (EAMT), pp. 261–268.

16. Tiedemann, Jörg. 2009. News from OPUS-A collection of multilingual parallel corpora with tools and interfaces. *In*: Recent Advances in Natural Language Processing, pp. 237–248.

17. Marasek, Krzysztof. 2012. TED Polish-to-English translation system for the IWSLT 2012. *In*: IWSLT, pp. 126–129.

18. Wołk, Krzysztof and Marasek, Krzysztof. 2013. Polish–english speech statistical machine translation systems for the IWSLT 2013. *In*: Proceedings of the 10th International Workshop on Spoken Language Translation, Heidelberg, Germany, pp. 113–119.

19. Berrotarán, Gonzalo Garcia, Carrascosa, Rafael and Vine, Andrew. 2015. Yalign documentation, http://yalign.readthedocs.org/en/latest/, retrieved on June 17, 2015.

20. Musso, Gabe. 2015. Sequence alignment (Needleman-Wunsch, Smith-Waterman). Available: http://www.cs.utoronto.ca/~brudno/bcb410/lec2notes.pdf, retrieved on June 17, 2015.

21. Joachims, Thorsten. 1998. Text categorization with support vector machines: Learning with many relevant features. Springer Berlin Heidelberg, pp. 137–142.

22. Wołk, Krzysztof and Marasek, Krzysztof. 2014. Real-time statistical speech translation. *In*: New Perspectives in Information Systems and Technologies, Volume 1. Springer International Publishing, pp. 107–113.

23. Varga, Dániel et al. 2007. Parallel corpora for medium density languages. Amsterdam studies in the theory and history of linguistic science series 4. 292: 247.

24. Wołk, Krzysztof and Marasek, Krzysztof. 2014. A sentence meaning based alignment method for parallel text corpora preparation. *In*: New Perspectives in Information Systems and Technologies, Volume 1. Springer International Publishing, pp. 229–237.

25. Hovy, Eduard. 1999. Toward finely differentiated evaluation metrics for machine translation. *In*: Proceedings of the EAGLES Workshop on Standards and Evaluation. Pisa, Italy.

26. Reeder, F. 2001. Additional mt-eval references. International Standards for Language Engineering, Evaluation Working Group.

27. Papineni, Kishore et al. 2002. BLEU: A method for automatic evaluation of machine translation. *In*: Proceedings of the 40th annual meeting on association for computational linguistics. Association for Computational Linguistics, pp. 311–318.

28. Axelrod, Amittai. 2006. Factored language model for statistical machine translation. M. Sc. Monograph University of Edinburgh.

29. Doddington, George. 2002. Automatic evaluation of machine translation quality using n-gram co-occurrence statistics. *In*: Proceedings of the second international conference on Human Language Technology Research. Morgan Kaufmann Publishers Inc., pp. 138–145.

30. Olive, Joseph. 2005. Global autonomous language exploitation (GALE). DARPA/IPTO Proposer Information Pamphlet.

31. Banerjee, Satanjeev and Lavie, Alon. 2005. Meteor: An automatic metric for MT evaluation with improved correlation with human judgments. *In*: Proceedings of the ACL Workshop on Intrinsic and Extrinsic Evaluation Measures for Machine Translation and/or Summarization, pp. 65–72.

32. Goodman, Joshua and Chen, Stanley. 1998. An empirical study of smoothing techniques for language modeling. Computer Science Group, Harvard University,
33. Perplexity [Online]. Hidden Markov Model Toolkit website. Cambridge University Engineering Dept. Available: http://www1.icsi.berkeley.edu/Speech/docs/HTKBook3.2/node188_mn.html, retrieved on November 29, 2015.
34. Isozaki, Hideki et al. 2010. Automatic evaluation of translation quality for distant language pairs. *In*: Proceedings of the 2010 Conference on Empirical Methods in Natural Language Processing. Association for Computational Linguistics, pp. 944–952.
35. European Medicines Agency (EMEA) [Online]. Available: http://opus.lingfil.uu.se/EMEA.php, retrieved on August 7, 2013.
36. European Parliament Proceedings (Europarl) [Online]. Available: http://www.statmt.org/europarl/, retrieved on August 7, 2013.
37. Jinseok, Kim. Lambda N. Gamma [Online]. 2004. Available: http://www.utexas.edu/courses/schwab/sw318_spring_2004/SolvingProblems/Class11_LambdaNGamma.ppt, retrieved on August 17, 2014.
38. Gautheir, Thomas D. 2001. Detecting trends using spearman's rank correlation coefficient. Environmental Forensics 2: 359–362.
39. Deng, Yonggang, Kumar, Shankar and Byrne, William. 2007. Segmentation and alignment of parallel text for statistical machine translation. Natural Language Engineering 13: 235–260.
40. Braune, Fabienn and Fraser, Alexander. 2010. Improved unsupervised sentence alignment for symmetrical and asymmetrical parallel corpora. *In*: Proceedings of the 23rd International Conference on Computational Linguistics: Posters. Association for Computational Linguistics, pp. 81–89.
41. Santos, André. 2011. A survey on parallel corpora alignment. MI-STAR, pp. 117–128.
42. Gale, William A. and Church, Kenneth Ward. 1991. Identifying word correspondences in parallel texts. *In*: HLT, pp. 152–157.
43. Brown, Peter F., Lai, Jennifer C. and Mercer, Robert L. 1991. Aligning sentences in parallel corpora. *In*: Proceedings of the 29th annual meeting on Association for Computational Linguistics. Association for Computational Linguistics, pp. 169–176.
44. Filtering and aligning tool. Available: http://wolk.pl/tool-for-parallel-corpora-filtering-and-aligning/, retrieved on June 10, 2015.
45. Cettolo, Mauro, Bertoldi, Nicola and Federico, Marcello. 2011. Methods for smoothing the optimizer instability in SMT. MT Summit XIII: The Thirteenth Machine Translation Summit, pp. 32–39.
46. Snover, Matthew et al. 2006. A study of translation edit rate with targeted human annotation. *In*: Proceedings of Association for Machine Translation in the Americas, pp. 223–231.
47. Costa-Jussà, Marta R. and Fonollosa, José A.R. 2010. Using linear interpolation and weighted reordering hypotheses in the moses system. *In*: LREC.
48. International Workshop on Spoken Language Translation (IWSLT). Available: http://www.iwslt2013.org/, retrieved on August 7, 2013.
49. Abbyy Aligner. Available: http://www.abbyy.com/aligner/, retrieved on August 7, 2013.
50. Unitex/Gramlab, Available: http://www-igm.univ-mlv.fr/~unitex, retrieved on August 7, 2013.
51. Hunalign–sentence aligner, Available: http://mokk.bme.hu/resources/hunalign/, retrieved on August 8, 2013.
52. Bleualign, https://github.com/rsennrich/Bleualign, retrieved on August 8, 2013.
53. Paumier, Sébastien, Nakamura, Takuya and Voyatzi, Stavroula. 2009. UNITEX, a Corpus Processing System with Multi-Lingual Linguistic Resources. eLEX2009, pp. 173.
54. Bonhomme, Patrice and Romary, Laurent. 1995. The lingua parallel concordancing project: Managing multilingual texts for educational purpose. Proceedings of Language Engineering, 95: 26–30.

55. Hsu, Bo-June and Glass, James. 2008. Iterative language model estimation: Efficient data structure & algorithms. *In*: Proceedings of Interspeech, pp. 1–4.
56. Proceedings of International Workshop on Spoken Language Translation IWSLT 2014, Tahoe Lake, USA, 2014. Available Online: http://workshop2014.iwslt.org, retrieved on August 8, 2013.
57. Bisazza, Arianna et al. 2011. Fill-up versus interpolation methods for phrase-based SMT adaptation. *In*: IWSLT, pp. 136–143.
58. Bojar, Ondrej. 2011. Rich morphology and what can we expect from hybrid approaches to MT. Invited talk at International Workshop on Using Linguistic Information for Hybrid Machine Translation LIHMT-2011.
59. Graliński, Filip, Jassem, Krzysztof and Junczys-Dowmunt, Marcin. 2013. PSI-toolkit: A natural language processing pipeline. *In*: Computational Linguistics. Springer Berlin Heidelberg, pp. 27–39.
60. Wołk, Krzysztof and Marasek, Krzysztof. 2015. Polish-english statistical machine translation of medical texts. *In*: New Research in Multimedia and Internet Systems. Springer International Publishing, pp. 169–179.
61. Chiang, David. 2007. Hierarchical phrase-based translation. Computational Linguistics, 33.2: 201–228.
62. Durrani, Nadir, Schmid, Helmut and Fraser, Alexander. 2011. A joint sequence translation model with integrated reordering. *In*: Proceedings of the 49th Annual Meeting of the Association for Computational Linguistics: Human Language Technologies-Volume 1. Association for Computational Linguistics, pp. 1045–1054.
63. Goeuriot, Lorraine et al. 2012. Report on and prototype of the translation support. Khresmoi Public Deliverable, p. 3.
64. Pletneva, Natalia and Vargas, Alejandro. 2011. D8. 1.1. Requirements for the general public health search. Khresmoi Project public deliverable.
65. Gschwandtner, M., Kritz, M. and Boyer, C. 2011. Requirements of the health professional research. Technical Report.
66. GGH Benefits. Medical phrases and terms translation demo. 2014, retrieved on August 7, 2013.
67. Karliner, Leah S. et al. 2007. Do professional interpreters improve clinical care for patients with limited English proficiency? A systematic review of the literature. Health Services Research 42.2: 727–754.
68. Randhawa, Gurdeeshpal et al. 2013. Using machine translation in clinical practice. Canadian Family Physica 59: 382–383.
69. Deschenes, S. 5 benefits of healthcare translation technology [Online]. Healthcare Finance News. October 16, 2012. Available: http://www.healthcarefinancenews.com/news/5-benefits-healthcare-translation-technology, retrieved on June 10, 2015.
70. Zadon, Cruuz. Man vs. machine: The benefits of medical translation services [Online]. Ezine Articles: Healthcare Systems, July 25, 2013. Available: http://ezinearticles.com/?Man-Vs-Machine:-The-Benefits-of-Medical-Translation-Services&id=7890538, retrieved on August 7, 2013.
71. Durrani, Nadir et al. 2013. Munich-Edinburgh-Stuttgart submissions of OSM systems at WMT13. *In*: Proceedings of the Eighth Workshop on Statistical Machine Translation, pp. 120–125.
72. Koehn, Philipp and Hoang, Hieu. 2007. Factored translation models. Proceedings of the 2007 Joint Conference on Empirical Methods in Natural Language Processing and Computational Natural Language Learning EMNLP-CoNLL, pp. 868–876.
73. Bikel, Daniel M. 2004. Intricacies of Collins' parsing model. Computational Linguistics 30: 479–511.
74. Dyer, Chris, Chahuneau, Victor and Smith, Noah A. 2013. A simple, fast, and effective reparameterization of IBM Model 2. Association for Computational Linguistics.

75. Bojar, Ondrej et al. 2014. Findings of the 2014 workshop on statistical machine translation. *In*: Proceedings of the Ninth Workshop on Statistical Machine Translation. Association for Computational Linguistics, Baltimore, MD, USA, pp. 12–58.

76. Hasan, A., Islam, Saria and Rahman, M. 2012. A comparative study of Witten Bell and Kneser-Ney smoothing methods for statistical machine translation. Journal of Information Technology 1: 1–6.

77. Radziszewski, Adam and Sniatowski, Tomasz. 2011. Maca-a configurable tool to integrate Polish morphological data [Online]. *In*: Proceedings of the Second International Workshop on Free/Open-Source Rule-Based Machine Translation (2011: Barcelona). Available: http://hdl. handle. net/10609/5645.

78. Bahdanau, Dzmitry, Cho, Kyunghyun and Bengio, Yoshua. 2014. Neural Machine Translation by Jointly Learning to Align and Translate. arXiv preprint arXiv:1409.0473.

79. Durrani, Nadir et al. 2014. Investigating the usefulness of generalized word representations in SMT. *In*: Proceedings of the 25th Annual Conference on Computational Linguistics (COLING), Dublin, Ireland, pp. 421–432.

80. Durrani, Nadir et al. 2014. Integrating an unsupervised transliteration model into statistical machine translation. Proceedings of the 14th Conference of the European Chapter of the Association for Computational Linguistics, EACL, pp. 148–153.

81. Wołk, Krzysztof and Marasek, Krzysztof. 2013. Alignment of the polish-english parallel text for a statistical machine translation. Computer Technology and Application 4. David Publishing, ISSN:1934-7332 (Print), ISSN: 1934-7340 (Online), pp. 575–583.

82. Koehn, Philipp and Haddow, Barry. 2012. Towards effective use of training data in statistical machine translation. *In*: Proceedings of the Seventh Workshop on Statistical Machine Translation. Association for Computational Linguistics, pp. 317–321.

83. Clark, Jonathan H. et al. 2011. Better hypomonograph testing for statistical machine translation: Controlling for optimizer instability. *In*: Proceedings of the 49th Annual Meeting of the Association for Computational Linguistics: Human Language Technologies: Short monographs-Volume 2. Association for Computational Linguistics, pp. 176–181.

84. Durrani, Nadir et al. 2013. Edinburgh's machine translation systems for European language pairs. *In*: Proceedings of the Eighth Workshop on Statistical Machine Translation, pp. 114–121.

85. Koehn, Philipp. 2009. Statistical machine translation. Cambridge University Press.

86. Askarieh, Sona. 2014. Cohesion and Comprehensibility in Swedish-English Machine Translated Texts.

87. Wojak, Aleksandra and Gralinski, Filip. 2010. Matura evaluation experiment based on human evaluation of machine translation. *In*: IMCSIT, pp. 547–551.

88. Lopez, Adam. 2008. Statistical machine translation. ACM Computing Surveys (CSUR) 40: 8.

89. Hutchins, John. 2003. Has machine translation improved? Some historical comparisons. *In*: Proceedings of the 9th MT Summit, pp. 181–188.

90. Hutchins, W. John. 2003. Machine translation: half a century of research and use. Proc. UNED Summer School, pp. 1–24.

91. Agarwal, Abhaya and Lavie, Alon. 2008. Meteor, M-BLEU and M-TER: Evaluation metrics for high-correlation with human rankings of machine translation output. *In*: Proceedings of the Third Workshop on Statistical Machine Translation. Association for Computational Linguistics, pp. 115–118.

92. Dugast, Loïc, Senellart, Jean and Koehn, Philipp. 2007. Statistical post-editing on systran's rule-based translation system. *In*: Proceedings of the Second Workshop on Statistical Machine Translation. Association for Computational Linguistics, pp. 220–223.

93. Zhang, Ruiqiang, et al. 2004. A unified approach in speech-to-speech translation: Integrating features of speech recognition and machine translation. *In*: Proceedings of the 20th international conference on Computational Linguistics. Association for Computational Linguistics, p. 1168.

94. Matusov, Evgeny, Ueffing, Nicola and Ney, Hermann. 2006. Computing consensus translation for multiple machine translation systems using enhanced hypomonograph alignment. Proceedings of the 11st Conference of the European Chapter of the Association for Computational Linguistics (EACL), pp. 33–40.

95. Weiss, Sandra and Ahrenberg, Lars. 2012. Error profiling for evaluation of machine-translated text: A Polish-English case study. *In*: LREC, pp. 1764–1770.

96. Dowty, David R. 1986. The effects of aspectual class on the temporal structure of discourse: Semantics or pragmatics? Linguistics and Philosophy 9: 37–61.

97. Kupść, Anna. 2003. Aspect assignment in a knowledge-based english-polish machine translation system. *In*: Intelligent Information Processing and Web Mining. Springer Berlin Heidelberg, pp. 159–167.

98. Aker, Ahmet, Kanoulas, Evangelos and Gaizauskas, Robert J. 2012. A light way to collect comparable corpora from the Web. *In*: LREC, pp. 15–20.

99. Strötgen, Jannik, Gertz, Michael and Junghans, Conny. 2011. An event-centric model for multilingual document similarity. *In*: Proceedings of the 34th international ACM SIGIR conference on Research and development in Information Retrieval. ACM, pp. 953–962.

100. Paramita, Monica Lestari, et al. 2013. Methods for collection and evaluation of comparable documents. *In*: Building and Using Comparable Corpora. Springer Berlin Heidelberg, pp. 93–112.

101. Volk, Martin and Harder, Søren. 2007. Evaluating MT with translations or translators. What is the difference. Machine Translation Summit XI Proceedings.

102. Hildebrand, Almut Silja, et al. 2005. Adaptation of the translation model for statistical machine translation based on information retrieval. Proceedings of the 10th Conference of the European Association for Machine Translation EAMT, pp. 133–142.

103. Yasuda, Keiji, et al. 2008. Method of selecting training data to build a compact and efficient translation model. *In*: IJCNLP, pp. 655–660.

104. Sethy, Abhinav, Georgiou, Panayiotis G. and Narayanan, Shrikanth. 2006. Selecting relevant text subsets from web-data for building topic specific language models. *In*: Proceedings of the Human Language Technology Conference of the NAACL, Companion Volume: Short Monographs. Association for Computational Linguistics, pp. 145–148.

105. Lü, Yajuan, Huang, Jin and Liu, Qun. 2007. Improving statistical machine translation performance by training data selection and optimization. *In*: EMNLP-CoNLL, pp. 343–350.

106. Foster, George and Kuhn, Roland. 2007. Mixture-model adaptation for SMT. Proceedings of the Second Workshop on Statistical Machine Translation.

107. Civera, Jorge and Juan, Alfons. 2007. Domain adaptation in statistical machine translation with mixture modelling. *In*: Proceedings of the Second Workshop on Statistical Machine Translation. Association for Computational Linguistics, pp. 177–180.

108. Koehn, Philipp and Schroeder, Josh. 2007. Experiments in domain adaptation for statistical machine translation. *In*: Proceedings of the Second Workshop on Statistical Machine Translation. Association for Computational Linguistics, pp. 224–227.

109. Xu, Jia, et al. 2007. Domain dependent statistical machine translation. Proceedings of the MT Summit XI.

110. Finch, Andrew and Sumita, Eiichiro. 2008. Dynamic model interpolation for statistical machine translation. *In*: Proceedings of the Third Workshop on Statistical Machine Translation. Association for Computational Linguistics, pp. 208–215.

111. Wu, Hua, Wang, Haifeng and Zong, Chengqing. 2008. Domain adaptation for statistical machine translation with domain dictionary and monolingual corpora. *In*: Proceedings of the 22nd International Conference on Computational Linguistics-Volume 1. Association for Computational Linguistics, pp. 993–1000.
112. Cheng, Pu-Jen, et al. 2004. Creating multilingual translation lexicons with regional variations using web corpora. *In*: Proceedings of the 42nd Annual Meeting on Association for Computational Linguistics. Association for Computational Linguistics, p. 534.
113. Axelrod, Amittai, He, Xiaodong and Gao, Jianfeng. 2011. Domain adaptation via pseudo in-domain data selection. *In*: Proceedings of the Conference on Empirical Methods in Natural Language Processing. Association for Computational Linguistics, pp. 355–362.
114. Tillmann, Christoph. 2004. A unigram orientation model for statistical machine translation. *In*: Proceedings of HLT-NAACL 2004: Short Monographs. Association for Computational Linguistics, pp. 101–104.
115. Galley, Michel and Manning, Christopher D. 2008. A simple and effective hierarchical phrase reordering model. *In*: Proceedings of the Conference on Empirical Methods in Natural Language Processing. Association for Computational Linguistics, pp. 848–856.
116. Och, Franz Josef and Ney, Hermann. 2003. A systematic comparison of various statistical alignment models. Computational Linguistics 29: 19–51.
117. Och, Franz Josef. 1995. Maximum-likelihood-schätzung von Wortkategorien mit verfahren der kombinatorischen optimierung. Studienarbeit, Friedrich-Alexander-Universität, Erlangen-Nürnberg, Germany.
118. Och, Franz Josef. 1999. An efficient method for determining bilingual word classes. *In*: Proceedings of the Ninth Conference on European Chapter of the Association for Computational Linguistics. Association for Computational Linguistics, pp. 71–76.
119. KantanMT—a sophisticated and powerful Machine Translation solution in an easy-to-use package [http://www.kantanmt.com].
120. Linguistic Intelligence Research Group, NTT Communication Science Laboratories. RIBES: Rank-based Intuitive Bilingual Evaluation Score [Online]. Available: http://www.kecl.ntt.co.jp/icl/lirg/ribes/, retrieved on August 7, 2013.
121. Chahuneau, Victor, Smith, Noah and Dyer, Chris. 2012. Pycdec: A python interface to cdec. The Prague Bulletin of Mathematical Linguistics 98: 51–61.
122. Dyer, Chris et al. 2010. cdec: A decoder, alignment, and learning framework for finite-state and context-free translation models. *In*: Proceedings of the ACL 2010 System Demonstrations. Association for Computational Linguistics, pp. 7–12.
123. Quirk, Chris and Menezes, Arul. 2006. Do we need phrases? Challenging the conventional wisdom in statistical machine translation. *In*: Proceedings of the main Conference on Human Language Technology Conference of the North American Chapter of the Association of Computational Linguistics. Association for Computational Linguistics, pp. 9–16.
124. Marino, José B. et al. 2006. N-gram-based machine translation. Computational Linguistic 32s: 527–549.
125. Costa-jussà, Marta R. et al. 2007. Analysis and system combination of phrase-and n-gram-based statistical machine translation systems. *In*: Human Language Technologies 2007: The Conference of the North American Chapter of the Association for Computational Linguistics; Companion Volume, Short Monographs. Association for Computational Linguistics, pp. 137–140.
126. Johnson, John Howard et al. 2007. Improving translation quality by discarding most of the phrasetable. Proceedings of the 2007 Joint Conference on Empirical Methods in Natural Language Processing and Computational Natural Language Learning (EMNLP-CoNLL).

127. Wu, Hua and Wang, Haifeng. 2007. Comparative study of word alignment heuristics and phrase-based SMT. Proceedings of the MT Summit XI.

128. Kutsumi, Takeshi et al. 2005. Selection of entries for a bilingual dictionary from aligned translation equivalents using support vector machines. *In*: Proceedings of Pacific Association for Computational Linguistics.

129. Eck, Matthias, Vogel, Stephan and Waibel, Alex. 2007. Translation model pruning via usage statistics for statistical machine translation. *In*: Human Language Technologies 2007: The Conference of the North American Chapter of the Association for Computational Linguistics; Companion Volume, Short Monographs. Association for Computational Linguistics, pp. 21–24.

130. Eck, Stephen Vogal Matthias and Waibel, Alex. 2007. Estimating phrase pair relevance for translation model pruning. MT Summit XI.

131. Durrani, Nadir and Koehn, Philipp. 2014. Improving machine translation via triangulation and transliteration. *In*: Proceedings of the 17th Annual Conference of the European Association for Machine Translation (EAMT), Dubrovnik, Croatia.

132. Durrani, Nadir et al. 2014. Investigating the usefulness of generalized word representations in SMT. *In*: Proceedings of the 25th Annual Conference on Computational Linguistics (COLING), Dublin, Ireland, pp. 421–432.

133. Aiken, Milam and Balan, Shilpa. 2011. An analysis of Google Translate accuracy. Translation Journal 16: 1–3.

134. Wesley-Tanaskovic, Ines, Tocatlian, Jacques and Roberts, Kenneth H. 1994. Expanding access to science and technology: The role of information technologies. Proceedings of the Second International Symposium on the Frontiers of Science and Technology Held in Kyoto, Japan, 12–14 May 1992. United Nations University Press.

135. Helft, Miguel. 2010. Google's computing power refines translation tool. The New York Times.

136. Lehrberger, John and Bourbeau, Laurent. 1988. Machine translation: Linguistic characteristics of MT systems and general methodology of evaluation. John Benjamins Publishing.

137. Costa-Jussa, Marta R. et al. 2012. Study and comparison of rule-based and statistical catalan-spanish machine translation systems. Computing and Informatics 31: 245–270.

138. Weiss, Sandra. 2011. Cohesion and Comprehensibility in Polish-English Machine Translated Texts.

139. Costa-Jussà, Marta R., Farrús, Mireia and Pons, Jordi Serrano. 2012. Machine translation in medicine. Proceedings in ARSA-Advanced Research in Scientific Areas, p. 1.

140. Smith, Ray. 2007. An overview of the Tesseract OCR engine. *In*: icdar. IEEE, pp. 629–633.

141. Jurafsky, Daniel and Martin, James H. 2000. Speech and language processing: An introduction to natural language processing. Computational Linguistics, and Speech Recognition, Pearson Education India.

142. Maccartney, Bill. 2005. NLP lunch tutorial: Smoothing.

143. Tian, Liang, Wong, Fai and Chao, Sam. 2011. Word alignment using GIZA++ on Windows. Machine Translation.

144. Lavie, Alon. 2010. Evaluating the output of machine translation systems. AMTA Tutorial.

145. Jassem, Krzysztof. 1997. Poleng–A machine translation system based on an electronic dictionary. Speech and Language Technology 1: 161–194.

146. Jassem, Krzysztof, Graliński, Filip and Krynicki, Grzegorz. 2000. Poleng-adjusting a rule-based polish-english machine translation system by means of corpus analysis. *In*: Proceedings of the 5th European Association for Machine Translation Conference, pp. 75–82.

147. Junczys-Dowmunt Marcin. 2011. A comparison of search algorithms for syntax-based SMT, Speech, Language and Technology 01/2011.
148. Bojar, Ondrej, Rosa, Rudolf and Tamchyna, Aleš. 2013. Chimera–three heads for English-to-Czech translation. *In*: Proceedings of the Eighth Workshop on Statistical Machine Translation. Association for Computational Linguistics Sofia, Bulgaria, pp. 90–96.
149. Wołk, Krzysztof and Marasek, Krzysztof. 2014. Enhanced bilingual evaluation understudy. Lecture Notes on Information Theory Vol 2.2.
150. Wang, Chao, Collins, Michael and Koehn, Philipp. 2007. Chinese syntactic reordering for statistical machine translation. *In*: EMNLP-CoNLL, pp. 737–745.
151. Zhang, Ruiqiang et al. 2004. A unified approach in speech-to-speech translation: Integrating features of speech recognition and machine translation. *In*: Proceedings of the 20th international conference on Computational Linguistics. Association for Computational Linguistics, p. 1168.
152. Wu, Xianchao et al. 2011. Extracting pre-ordering rules from predicate-argument structures. *In*: IJCNLP, pp. 29–37.
153. Fragoso, Victor et al. 2011. Translatar: A mobile augmented reality translator. *In*: Applications of Computer Vision (WACV), 2011 IEEE Workshop on. IEEE, pp. 497–502.
154. Cettolo, Mauro et al. 2014. Report on the 11th IWSLT Evaluation Campaign, IWSLT 2014. *In*: Proceedings of the Eleventh International Workshop on Spoken Language Translation (IWSLT), Lake Tahoe, CA, pp. 2–17.
155. Zeng, Wei and Church, Richard L. 2009. Finding shortest paths on real road networks: The case for A*. International Journal of Geographical Information Science 23: 531–543.
156. Daniels, Peter T. and Bright, William. 1996. The world's writing systems. Oxford University Press.
157. Coleman, John. 2014. A Speech Is Not an Essay, Harvard Business Review.
158. Ager, Simon. 2013. Differences between writing and speech [Online]. Omniglot—the online encyclopedia of writing systems and languages. Available: http://www.omniglot.com/writing/writingvspeech.htm, retrieved on August 8, 2013.
159. Ruiz, Nicholas and Federico, Marcello. 2014. Complexity of spoken versus written language for machine translation. *In*: Proceedings of the 17th Annual Conference of the European Association for Machine Translation, pp. 173–180.
160. Swan, Oscar E. 2003. Polish Grammar in a Nutshell [Online]. 2003. Available: http://www.skwierzyna.net/polishgrammar.pdf, retrieved on 18.04.2015.
161. Bojar, Ondřej, Matusov, Evgeny and Ney, Hermann. 2006. Czech-English phrase-based machine translation. *In*: Advances in Natural Language Processing. Springer Berlin Heidelberg, pp. 214–224.
162. Bojar, Ondřej and Hajič, Jan. 2008. Phrase-based and deep syntactic English-to-Czech statistical machine translation. *In*: Proceedings of the third Workshop on Statistical Machine translation. Association for Computational Linguistics, pp. 143–146.
163. Bojar, Ondřej. 2007. English-to-Czech factored machine translation. *In*: Proceedings of the Second Workshop on Statistical Machine Translation. Association for Computational Linguistics, pp. 232–239.
164. Huet, Stéphane, Manishina, Elena and Lefèvre, Fabrice. 2013. Factored machine translation systems for Russian-English. *In*: Proceedings of the Eighth Workshop on Statistical Machine Translation, pp. 152–155.
165. Bojar, Ondřej and Zeman, Daniel. 2014. Czech machine translation in the project CzechMATE. The Prague Bulletin of Mathematical Linguistics 101: 71–96.
166. Kolovratnik, David, Klyueva, Natalia and Bojar, Ondrej. 2009. Statistical machine translation between related and unrelated languages. *In*: Proceedings of the Conference on Theory and Practice on Information Technologies, pp. 31–36.

167. Steinberger, Ralf et al. 2013. Dgt-tm: A freely available translation memory in 22 languages. arXiv preprint arXiv:1309, 5226.
168. Jassem, Krzysztof. 2004. Applying Oxford-PWN english-polish dictionary to machine translation. *In*: Proceedings of the 9th European Association for Machine Translation Workshop on Broadening Horizons of Machine Translation and its Applications.
169. Besacier, Laurent et al. 2014. Word confidence estimation for speech translation. *In*: International Workshop on Spoken Language Translation.
170. Wołk, Krzysztof and Marasek, Krzysztof. 2014. Building subject-aligned comparable corpora and mining it for truly parallel sentence pairs. Procedia Technology 18: 126–132.
171. Moore, Robert C. and Quirk, Chris. 2009. Improved smoothing for N-gram language models based on ordinary counts. *In*: Proceedings of the ACL-IJCNLP 2009 Conference Short Monographs. Association for Computational Linguistics, pp. 349–352.
172. Hsu, Bo-June and Glass, James. 2008. Iterative language model estimation: Efficient data structure & algorithms. *In*: Proceedings of Interspeech, pp. 1–4.
173. Kirchhoff, Katrin et al. 2011. Application of statistical machine translation to public health information: A feasibility study. Journal of the American Medical Informatics Association 18.4: 473–478.
174. Malouf, Robert. 2002. A comparison of algorithms for maximum entropy parameter estimation. *In*: Proceedings of the 6th conference on Natural language learning-Volume 20. Association for Computational Linguistics, pp. 1–7.
175. Rottmann, Kay and Vogel, Stephan. 2007. Word reordering in statistical machine translation with a POS-based distortion model. Proc. of TMI, pp. 171–180.
176. Borman, Sean. 2004. The expectation maximization algorithm-a short tutorial. Submitted for Publication, pp. 1–9.
177. Collins, Michael. 2011. Statistical machine translation: IBM models 1 and 2. Columbia: Columbia University.
178. Och, Franz Josef and Ney, Hermann. 2003. A systematic comparison of various statistical alignment models. Computational Linguistics 29: 19–51.
179. Vulić, I., Term Alignment: State of the Art Overview [Online]. Katholieke Universiteit Leuven, 2010, Available: http://people.cs.kuleuven.be/~ivan.vulic/Files/TASOA.pdf, retrieved on August 7, 2013.
180. Schoenemann, Thomas. 2010. Computing optimal alignments for the IBM-3 translation model. *In*: Proceedings of the Fourteenth Conference on Computational Natural Language Learning. Association for Computational Linguistics, pp. 98–106.
181. Fernández, Pablo Malvar. 2008. Improving word-to-word alignments using morphological information. 2008. PhD Monograph. San Diego State University.
182. Brown, Peter F. et al. 1993. The mathematics of statistical machine translation: Parameter estimation. Computational Linguistics 19: 263–311.
183. Knight, Kevin. 1999. A statistical MT tutorial workbook. Manuscript prepared for the 1999 JHU Summer Workshop.
184. Koehn, Philipp. 2010. Moses, statistical machine translation system, user manual and code guide.
185. Zhao, Shaojun and Gildea, Daniel. 2010. A fast fertility hidden Markov model for word alignment using MCMC. *In*: Proceedings of the 2010 Conference on Empirical Methods in Natural Language Processing. Association for Computational Linguistics, pp. 596–605.
186. Ma, Yanjun. 2009. Out of GIZA—Efficient Word Alignment Models for SMT. NCLT Seminar Series.
187. Federico, Marcello, Bertoldi, Nicola and Cettolo, Mauro. 2008. IRSTLM: An open source toolkit for handling large scale language models. *In*: Interspeech, pp. 1618–1621.

188. Snover, Matthew et al. 2006. A study of translation edit rate with targeted human annotation. *In*: Proceedings of association for machine translation in the Americas, pp. 223–231.
189. Denkowski, Michael and Lavie, Alon. 2011. Meteor 1.3: Automatic metric for reliable optimization and evaluation of machine translation systems. *In*: Proceedings of the Sixth Workshop on Statistical Machine Translation. Association for Computational Linguistics, pp. 85–91.
190. Jurafsky, Dan. 2015. Language modeling: Introduction to n-grams [Online]. Stanford University. Available: https://web.stanford.edu/class/cs124/lec/languagemodeling.pdf, retrieved on November 29, 2015.
191. Borisov, Alexey and Galinskaya, Irina. 2014. Yandex school of data analysis russian-english machine translation system for wmt14. ACL 2014, p. 66.
192. Michie, Donald, Spiegelhalter, David J. and Taylor, Charles C. 1994. Machine learning, neural and statistical classification.
193. Abdelali, Ahmed et al. 2014. The AMARA corpus: Building parallel language resources for the educational domain. *In*: Proceedings of the 9th International Conference on Language Resources and Evaluation, Reykjavik, Iceland. May.
194. Li, Da and Becchi, Michela. 2012. Multiple pairwise sequence alignments with the needleman-wunsch algorithm on GPU. *In*: High Performance Computing, Networking, Storage and Analysis (SCC), 2012 SC Companion. IEEE, pp. 1471–1472.
195. Yang, Wei and Lepage, Yves. 2014. Inflating a training corpus for SMT by using unrelated unaligned monolingual data. *In*: Advances in Natural Language Processing. Springer International Publishing, pp. 236–248.
196. Bergroth, Lasse, Hakonen, Harri and Raita, Timo. 2000. A survey of longest common subsequence algorithms. *In*: String Processing and Information Retrieval, 2000. SPIRE 2000. Proceedings. Seventh International Symposium on. IEEE, pp. 39–48.
197. Sharp, Bernadette, Carl, Michael, Zock, Michael and Jakobsen Arnt Lykke (eds.). 2011. Human-Machine Interaction in Translation: Proceedings of the 8th International NLPCS Workshop. Samfundslitteratur.
198. Cherry, Colin and Foster, George. 2012. Batch tuning strategies for statistical machine translation. *In*: Proceedings of the 2012 Conference of the North American Chapter of the Association for Computational Linguistics: Human Language Technologies. Association for Computational Linguistics, pp. 427–436.
199. Sakti, Sakriani et al. 2013. A-STAR: Toward translating Asian spoken languages. Computer Speech & Language 27: 509–527.
200. Sakti, Sakriani et al. 2008. Development of Indonesian large vocabulary continuous speech recognition system within A-STAR project. *In*: IJCNLP, pp. 19–24.
201. Gale, William A. and Church, Kenneth W. 1993. A program for aligning sentences in bilingual corpora. Computational Linguistics 19: 75–102.
202. Oyeka, Ikewelugo Cyprian Anaene and Ebuh, Godday Uwawunkonye. 2012. Modified Wilcoxon signed-rank test. Open Journal of Statistics 2: 172.
203. Och, Franz Josef. 2003. Minimum error rate training in statistical machine translation. Proceedings of the 41st Annual Meeting on Association for Computational Linguistics, Volume 1. Association for Computational Linguistics.
204. Nakamura, S. et al. 2007. A-star: Asia speech translation consortium. *In*: Proc. ASJ Autumn Meeting, page to appear, Yamanashi, Japan.
205. Choong, Careemah. 2014. The difference between written and spoken english. Assignment Unit 1 A in fulfillment of Graduate Diploma in English.
206. Koehn, Philipp, Och, Franz Josef and Marcu, Daniel. 2003. Statistical phrase-based translation. *In*: Proceedings of the 2003 Conference of the North American Chapter of the Association for Computational Linguistics on Human Language Technology-Volume 1. Association for Computational Linguistics, pp. 48–54.

207. Frederick, Jelinek. 1997. Statistical methods for speech recognition. MIT Press.
208. Zespół Przetwarzania Sygnałów Agh. Ngram [Online]. Available: http://www.dsp.agh. edu.pl/doku.php?id=pl:resources:ngram, retrieved on August 7, 2015.
209. Hasan, A., Islam, Saria and Rahman, M. 2012. A comparative study of Witten Bell and Kneser-Ney smoothing methods for statistical machine translation. Journal of Information Technology 1: 1–6.
210. Bertoldi, Nicola, Haddow, Barry and Fouet, Jean-Baptiste. 2009. Improved minimum error rate training in Moses. The Prague Bulletin of Mathematical Linguistics 91: 7–16.
211. Collins, Michael. 1999. Head-driven statistical models for natural language parsing. PhD monograph. University of Pennsylvania.
212. Shannon, C.E. 1948. A Mathematical theory of communication. Bell System Technical Journal 27(3): 379–423.
213. Gerber, Laurie and Yang, Jin. 1997. Systran mt dictionary development. Machine Translation: Past, Present and Future. *In*: Proceedings of Machine Translation Summit VI. October.
214. Birch, Alexandra, Osborne, Miles and Koehn, Philipp. 2008. Predicting success in machine translation. Proceedings of the Conference on Empirical Methods in Natural Language Processing. Association for Computational Linguistics.
215. Grefenstette, Gregory and Tapanainen, Pasi. 1994. What is a word, what is a sentence? Problems of tokenisation. Rank Xerox Research Centre.
216. Trim, Craig. The Art Of Tokenization [Online]. Jan, 23, 2013. Available: https://www. ibm.com/developerworks/community/blogs/nlp/entry/tokenization?lang=en, retrieved on November 30, 2015.
217. Wieczorek, M. 2011. Linguistic aspects of text-to-speech synmonograph. Uniwersytet Im. Adama Mickiewicza, Poznań.
218. Wołk, Krzysztof, Rejmund, Emilia and Marasek, Krzysztof. 2015. Harvesting comparable corpora and mining them for equivalent bilingual sentences using statistical classification and analogy-based heuristics. *In*: Foundations of Intelligent Systems. Springer International Publishing, pp. 433–441.
219. Langa, Natalia and Wojak, Aleksandra. 2011. Ewaluacja systemów tłumaczenia automatycznego., Uniwersytet im. Adama Mickiewicza, Poznań.
220. Lo, Chi-Kiu and Wu, Dekai. 2011. MEANT: An inexpensive, high-accuracy, semi-automatic metric for evaluating translation utility via semantic frames. *In*: Proceedings of the 49th Annual Meeting of the Association for Computational Linguistics: Human Language Technologies-Volume 1. Association for Computational Linguistics, pp. 220–229.
221. Bojar, Ondřej and Wu, Dekai. 2012. Towards a predicate-argument evaluation for MT. *In*: Proceedings of the Sixth Workshop on Syntax, Semantics and Structure in Statistical Translation. Association for Computational Linguistics, pp. 30–38.
222. Pradhan, Sameer S. et al. 2004. Shallow semantic parsing using support vector machines. *In*: HLT-NAACL, pp. 233–240.
223. Tiedemann, Jörg. 2009. News from OPUS-A collection of multilingual parallel corpora with tools and interfaces. *In*: Recent Advances in Natural Language Processing, pp. 237–248.
224. Carmigniani, Julie. 2011. Augmented reality methods and algorithms for hearing augmentation. Florida Atlantic University.
225. NeWo LLC. "EnWo—English cam translator." Last modified January 12, 2014. https:// play.google.com/store/apps/details?id=com.newo.enwo.
226. Cui, Lei et al. 2013. Multi-domain adaptation for SMT using multi-task learning. *In*: EMNLP, pp. 1055–1065.
227. Google. "Translate API." Last modified April 6, 2015. https://cloud.google.com/ translate/v2/pricing.

228. Martedi, Sandy, Uchiyama, Hideaki and Saito, Hideo. 2010. Clickable augmented documents. *In*: 2010 IEEE International Workshop on Multimedia Signal Processing (MMSP). IEEE, pp. 162–166. https://ieeexplore.ieee.org/xpl/mostRecentIssue.jsp?punumber=5656074

229. Abbyy. "TextGrabber + Translator." Last modified February 16, 2015. http://abbyy-textgrabber.android.informer.com/.

230. Du, Jun et al. 2011. Snap and translate using windows phone. *In*: 2011 International Conference on Document Analysis and Recognition (ICDAR). IEEE, pp. 809–813. https://ieeexplore.ieee.org/xpl/mostRecentIssue.jsp?punumber=6065245

231. The Economist. Word Lens: This changes everything. Last modified December 18, 2010. http://www.economist.com/blogs/gulliver/2010/12/instant_translation.

232. Datamark, Inc. "OCR as a Real-Time Language Translator." Last modified July 30, 2012. https://www.datamark.net/blog/ocr-as-a-real-time-language-translator.

233. Khan, Tabish et al. 2014. Augmented reality based word translator. In International Journal of Innovative Research in Computer Science & Technology (IJIRCST), pp. 2(2).

234. Mahbub-Uz-Zaman, S. and Islam, Tanjina. 2012. Application of augmented reality: Mobile camera based bangla text detection and translation. 2012. PhD Thesis. BRAC University.

235. Emmanuel Ashish S. and Nithyanandam S. 2014. An optimal text recognition and translation system for smart phones using genetic programming and cloud. In International Journal of Engineering Science and Innovative Technology (IJESIT) 3(2): 437–443.

236. Toyama, Takumi et al. 2014. A mixed reality head-mounted text translation system using eye gaze input. *In*: Proceedings of the 19th international conference on Intelligent User Interfaces. ACM, pp. 329–334.

237. Information Systems Laboratory, Adam Mickiewicz University. "SyMGIZA++." Last modified April 27, 2011. http://psi.amu.edu.pl/en/index.php?title=SyMGIZA.

238. Dušek, Ondrej et al. 2014. Machine translation of medical texts in the khresmoi project. *In*: Proceedings of the Ninth Workshop on Statistical Machine Translation, pp. 221–228.

239. Worldwide, H. 1998. Medical Phrases and Terms Translation Demo.

240. Rojas, Raúl. 2013. Neural networks: A systematic introduction. Springer Science & Business Media.

241. Jesan, John Peter and Lauro, Donald M. 2003. Human brain and neural network behavior: A comparison. Ubiquity, 2003, November: 2–2.

242. Tiedemann, Jörg. 2009. News from OPUS-A collection of multilingual parallel corpora with tools and interfaces. *In*: Recent Advances in Natural Language Processing, pp. 237–248.

243. Cho, Kyunghyun et al. 2014. Learning phrase representations using RNN encoder-decoder for statistical machine translation. arXiv preprint arXiv:1406.1078, 2014.

244. Goodfellow, Ian J. et al. 2013. Maxout networks. arXiv preprint arXiv:1302.4389.

245. Pascanu, Razvan et al. 2013. How to construct deep recurrent neural networks. arXiv preprint arXiv:1312.6026.

246. Zeiler, Matthew D. 2012. Adadelta: An adaptive learning rate method. arXiv preprint arXiv:1212.5701.

247. Graves, Alex. 2012. Sequence transduction with recurrent neural networks. arXiv preprint arXiv:1211.3711.

248. Boulanger-Lewandowski, Nicolas, Bengio, Yoshua and Vincent, Pascal. 2013. Audio chord recognition with recurrent neural networks. *In*: ISMIR, pp. 335–340.

249. Mikolov, Tomas et al. 2011. Rnnlm-recurrent neural network language modeling toolkit. *In*: Proc. of the 2011 ASRU Workshop, pp. 196–201.

250. Roessler, Ross. 2010. A GPU implementation of needleman-wunsch, specifically for use in the program pyronoise 2. Computer Science & Engineering.

251. European Federation of Hard of Hearing People: State of subtitling access inEU. 2011 Report. http://ec:europa:eu/internal market/consultations/2011/audiovisual/non-registered-organisations/european-federation-of-hard-of-hearing-people-efhoh-en:pdf, 2011. [Online; accessed 30 Jan. 2016].
252. Romero-Fresco, Pablo and Pérez, Juan Martínez. 2015. Accuracy rate in live subtitling: The NER model. *In*: Audiovisual Translation in a Global Context. Palgrave Macmillan, London, pp. 28–50.
253. Dutka L., Szarkowska A., Chmiel A., Lijewska A., Krejtz K., Marasek K. and Brocki L. 2015. Are interpreters better respeakers? An exploratory study on respeaking competences. 5th International Symposium on respeaking, live subtitling and accessibility, Rome, 12 June. https://www.unint.eu/it/ricerca/gruppi-di-ricerca/8-pagina/494-respeaking-live-subtitling-and-accessibility.html
254. Hovy, Eduard H. 1999. Toward finely differentiated evaluation metrics for machine translation. *In*: Proceedings of the EAGLES Workshop on Standards and Evaluation. Pisa, Italy.
255. Wołk, Krzysztof and Koržinek, Danijel. 2016. Comparison and adaptation of automatic evaluation metrics for quality assessment of re-speaking. arXiv preprint arXiv:1601.02789.
256. Birch, Alexandra et al. 2013. The feasibility of HMEANT as a human MT evaluation metric. *In*: Proceedings of the Eighth Workshop on Statistical Machine Translation, pp. 52–61.
257. FUREY, Terrence S., et al. Support vector machine classification and validation of cancer tissue samples using microarray expression data. Bioinformatics, 2000, 16.10: 906–914.
258. Graves, Alex and Schmidhuber, Jürgen. 2005. Framewise phoneme classification with bidirectional LSTM and other neural network architectures. Neural Networks 18.5-6: 602–610.
259. Tai, Kai Sheng, Socher, Richard and Manning, Christopher D. 2015. Improved semantic representations from tree-structured long short-term memory networks. arXiv preprint arXiv:1503.00075.
260. US Department of Health and Human Services. 2012. PROMIS: Instrument Development and Psychometric Evaluation Scientific Standards.
261. Wołk, Krzysztof and Marasek, Krzysztof. 2015. Unsupervised comparable corpora preparation and exploration for bi-lingual translation equivalents. arXiv preprint arXiv:1512.01641.
262. Bonomi, A.E. et al. 1996. Multilingual translation of the Functional Assessment of Cancer Therapy (FACT) quality of life measurement system. Quality of Life Research 5.3: 309–320.
263. Wild, Diane et al. 2009. Multinational trials—recommendations on the translations required, approaches to using the same language in different countries, and the approaches to support pooling the data: The ISPOR patient-reported outcomes translation and linguistic validation good research practices task force report. Value in Health 12.4: 430–440.
264. Wołk, Krzysztof and Marasek, Krzysztof. 2015. Polish-English speech statistical machine translation systems for the IWSLT 2013. arXiv preprint arXiv:1509.09097.
265. Wang, Longyue et al. 2014. A systematic comparison of data selection criteria for smt domain adaptation. The Scientific World Journal, 2014.
266. Junczys-Dowmunt, Marcin and Szał, Arkadiusz. 2012. Symgiza++: Symmetrized word alignment models for statistical machine translation. *In*: Security and Intelligent Information Systems. Springer, Berlin, Heidelberg, pp. 379–390.
267. Moses statistical machine translation, "OOVs." 2015. http://www.statmt.org/moses/?n=Advanced.OOVs#ntoc2. Accessed 27 September 2015.

268. Moses statistical machine translation, "Build reordering model." 2013. http://www. statmt.org/moses/?n=FactoredTraining. Build Reordering Model. Accessed 10 October 2015.

269. Moore, Robert C. and Lewis, William. 2010. Intelligent selection of language model training data. *In*: Proceedings of the ACL 2010 conference short papers. Association for Computational Linguistics, pp. 220–224.

270. Dieny, Romain et al. 2011. Bioinformatics inspired algorithm for stereo correspondence. International Conference on Computer Vision Theory and Application, 465–473.

271. Robertson, Michael J. et al. 1989. The weighted index method: A new technique for analyzing planar optical waveguides. Journal of Lightwave Technology 7.12: 2105–2111.

272. Derksen, Shelley and Keselman, Harvey J. 1992. Backward, forward and stepwise automated subset selection algorithms: Frequency of obtaining authentic and noise variables. British Journal of Mathematical and Statistical Psychology 45.2: 265–282.

273. Wołk, Krzysztof, Marasek, Krzysztof and Wołk, Agnieszka. 2016. Exploration for Polish-* bi-lingual translation equivalents from comparable and quasi-comparable corpora. *In*: Computer Science and Information Systems (FedCSIS), 2016 Federated Conference on. IEEE, pp. 517–525.

274. Anderson, Stephen R. 2004. How many languages are there in the world. Linguistic Society of America.

275. List of languages by number of native speakers. 2016. Wikipedia https://en.wikipedia. org/wiki/List_of_languages_by_number_of_native_speakers. Accessed 16.02.2016.

276. Paolillo, John C. and Das, Anupam. 2006. Evaluating language statistics: The ethnologue and beyond. Contract report for UNESCO Institute for Statistics.

277. English language in Europe 2016. Wikipedia. https://en.wikipedia.org/wiki/English_ language_in_Europe. Accessed 16 February 2017.

278. Munteanu, Dragos Stefan, Fraser, Alexander and Marcu, Daniel. 2004. Improved machine translation performance via parallel sentence extraction from comparable corpora. *In*: Proceedings of the Human Language Technology Conference of the North American Chapter of the Association for Computational Linguistics: HLT-NAACL, 2004.

279. Callison-Burch, Chris and Osborne, Miles. 2002. Co-Training for statistical machine translation. 2002. PhD Thesis. Master's thesis, School of Informatics, University of Edinburgh.

280. Ueffing N., Haffari G. and Sarkar A. 2009. Semisupervised learning for machine translation. pp. 237–256. *In*: Cyril Goutte, Nicola Cancedda, Marc Dymetman and George Foster (eds.). Learning Machine Translation. MIT Press, Pittsburgh, pp. 237–256.

281. Mann, Gideon S. and Yarowsky, David. 2001. Multipath translation lexicon induction via bridge languages. *In*: Proceedings of the second meeting of the North American Chapter of the Association for Computational Linguistics on Language technologies. Association for Computational Linguistics, pp. 1–8.

282. Kumar, Shankar, Och, Franz J. and Macherey, Wolfgang. 2007. Improving word alignment with bridge languages. *In*: Proceedings of the 2007 Joint Conference on Empirical Methods in Natural Language Processing and Computational Natural Language Learning (EMNLP-CoNLL).

283. Wu, Hua and Wang, Haifeng. 2007. Pivot language approach for phrase-based statistical machine translation. Machine Translation 21.3: 165–181.

284. Habash, Nizar and Hu, Jun. 2009. Improving Arabic-Chinese statistical machine translation using English as pivot language. *In*: Proceedings of the Fourth Workshop on Statistical Machine Translation. Association for Computational Linguistics, pp. 173–181.

285. Eisele, Andreas et al. 2008. Hybrid machine translation architectures within and beyond the EuroMatrix project. *In*: Proceedings of the 12th Annual Conference of the European Association for Machine Translation (EAMT 2008), pp. 27–34.

286. Cohn, Trevor and Lapata, Mirella. 2007. Machine translation by triangulation: Making effective use of multi-parallel corpora. *In*: Proceedings of the 45th Annual Meeting of the Association of Computational Linguistics, pp. 728–735.

287. Leusch, Gregor et al. 2010. Multi-pivot translation by system combination. *In*: International Workshop on Spoken Language Translation (IWSLT 2010).

288. Bertoldi, Nicola et al. 2008. Phrase-based statistical machine translation with pivot languages. *In*: International Workshop on Spoken Language Translation (IWSLT).

289. Yujian, Li and Bo, Liu. 2007. A normalized Levenshtein distance metric. IEEE Transactions on Pattern Analysis and Machine Intelligence 29.6: 1091–1095.

290. Cao G., Nie J. and Bai, J. 2005. Integrating term relationships into language models. *In*: Proceedings of the 28th Annual International ACM SIGIR Conference on Research and Development in Information Retrieval. Salvador pp. 298–305.

291. Bellegarda, J. 2000. Data-driven semantic language modeling. *In*: Institute for Mathematics and its Applications Workshop.

292. Thomo, A. 2009. Latent semantic analysis (LSA) tutorial. http://webhome.cs.uvic.ca/~thomo/svd.pdf. Accessed 16 February 2007.

293. Verspoor, Karin et al. 2008. A semantics-enhanced language model for unsupervised word sense disambiguation. *In*: International Conference on Intelligent Text Processing and Computational Linguistics. Springer, Berlin, Heidelberg, pp. 287–298.

294. Lison, Pierre and Tiedemann, Jörg. Opensubtitles 2016: Extracting large parallel corpora from movie and tv subtitles. 2016.

295. Fujita, Atsushi and Isabelle, Pierre. 2015. Expanding paraphrase lexicons by exploiting lexical variants. *In*: Proceedings of the 2015 Conference of the North American Chapter of the Association for Computational Linguistics. Human Language Technologies, pp. 630–640.

296. Shen, Libin, Sarkar, Anoop and Och, Franz Josef. 2004. Discriminative reranking for machine translation. *In*: Proceedings of the Human Language Technology Conference of the North American Chapter of the Association for Computational Linguistics. HLT-NAACL.

297. Devlin, Jacob et al. 2014. Fast and robust neural network joint models for statistical machine translation. *In*: Proceedings of the 52nd Annual Meeting of the Association for Computational Linguistics (Volume 1: Long Papers), pp. 1370–1380.

298. Daumé Iii, Hal and Jagarlamudi, Jagadeesh. 2011. Domain adaptation for machine translation by mining unseen words. *In*: Proceedings of the 49th Annual Meeting of the Association for Computational Linguistics. Human Language Technologies: Short papers-Volume 2. Association for Computational Linguistics, pp. 407–412.

299. Lin, Sung-Chien et al. 1997. Chinese language model adaptation based on document classification and multiple domain-specific language models. *In*: Fifth European Conference on Speech Communication and Technology.

300. Gao, Jianfeng et al. 2002. Toward a unified approach to statistical language modeling for Chinese. ACM Transactions on Asian Language Information Processing (TALIP) 1.1: 3–33.

301. Koehn, Philipp. 2004. Pharaoh: A beam search decoder for phrase-based statistical machine translation models. *In*: Conference of the Association for Machine Translation in the Americas. Springer, Berlin, Heidelberg, pp. 115–124.

302. Gotoh, Yoshihiko and Renals, Steve. 2000. Sentence boundary detection in broadcast speech transcripts. *In*: ASR2000-Automatic Speech Recognition: Challenges for the new Millenium ISCA Tutorial and Research Workshop (ITRW).

303. Burnard, Lou. 1995. Users Reference Guide British National Corpus Version 1.0.

Index

Printed and bound by CPI Group (UK) Ltd, Croydon, CR0 4YY

24/10/2024

01778307-0007